Gene Regulation and Metabolism

Gene Regulation and Metabolism

Postgenomic Computational Approaches

edited by Julio Collado-Vides and Ralf Hofestädt

A Bradford Book
The MIT Press
Cambridge, Massachusetts
London, England

© 2002 Massachusetts Institute of Technology

All rights reserved. No part of this book may be reproduced in any form by any electronic or mechanical means (including photocopying, recording, or information storage and retrieval) without permission in writing from the publisher.

This book was set in Palatino on 3B2 by Asco Typesetters, Hong Kong and was printed and bound in the United States of America.

Library of Congress Cataloging-in-Publication Data

Gene regulation and metabolism : postgenomic computational approaches / edited by Julio Collado-Vides & Ralf Hofestädt.
    p.   cm. — (Computational molecular biology)
Includes bibliographical references and index.
ISBN 0-262-03297-X (hc. : alk. paper)
   1. Genetics—Mathematical models. 2. Molecular biology—Mathematical models.
I. Collado-Vides, Julio. II. Hofestädt, Ralf. III. Series.
QH438.4.M3 G46 2002
572.8′01′5118—dc21   2001056247

Contents

Preface

We are in the middle of a genome period marked by the full sequencing of complete genomes. Last year (2001) will be identified in the history of biology by the publication of the first draft of the complete sequence of the human genome. Much work still lies ahead to achieve the goal of fully finishing many of these eukaryotic and prokaryotic genomes that, as published, still contain gaps.

At a first glance, genomics has not produced a strong conceptual change in biology. The fundamental problems remain: understanding the origin of life, the complex organization of a cell, the pathways of differentiation, aging, and the molecular and cellular bases for the capabilities of the brain. What has happened is an explosion of molecular information; genomic sequences will be followed in the near future by exhaustive catalogs of protein interactions and protein function (as proteomics takes the lead). This wealth of information can be analyzed, visualized, and manipulated only with the help of computers. This basic contribution of computers was initially not recognized by biologists. Certainly, by the time of the beginning of GenBank, in the 1980s, the experimentalist could imagine an institute where computational biology was merely technical support for databases and access to GenBank, and maybe a classic Bohering metabolic chart hung on the wall (initiated in the 1960s by G. Michal). The influence of genomes is such that today what François Jacob conceived as the Mouse Institute would do much better having on staff experimentalists, computer scientists, statisticians, mathematicians, and computational biologists. We have reached a point where biology articles are published with contributions from researchers who recently were, for instance, computer scientists working in logic programming.

This is no small change if we remember the place of theoretical and mathematical biology as an activity that could be fascinating, but to a large extent was done in isolation, having little influence on mainstream experimental molecular biology. Today, the student, postdoctoral fellow, or even young professor who is knowledgeable both in biology and in computer science has much broader opportunities. Genomics may really be opening the door to a more profound conceptual change in the way we study living systems in the laboratory.

With a foot in sequence analysis, this book is centered on current computational approaches to metabolism and gene regulation. This is an area of computational biology that welcomes new methods, ideas, and approaches with the goal of generating a better understanding of the complex networks of metabolic and regulatory capabilities of the cell. Classical concepts have to be redefined or clarified to address the study of the genetics of populations and of the biochemical interactions and regulatory networks organizing a living system. Given the constant and pervading importance of comparative genomics, these concepts must be precise when comparing genes, proteins, and systems across different species.

The first chapter, by Jeremy Ahouse, is an exercise in thinking about the concept of homology (the common origin of similarities) in order to use it adequately when considering homologous networks of gene regulation between species.

Currently, DNA sequence data is the most abundant material with which to begin a project in computational biology. Raw sequences from genomes have to be analyzed and annotated, in ways that improve continuously as the databases expand and sharper methods are used. The second chapter, by Rolf Apweiler and colleagues, describes an integrated system for this task. Databases centering on specific signals, motifs, or structures have exploded in number in the last years. The databases describe those pieces of macromolecules whose function we know, and therefore are essential for algorithmic analyses. The third chapter, by the team of Ralf Hofestädt, shows a system capable of integrating data from different databases, and its subsequent use in the integration and modeling of metabolic pathways using a rule-based system.

Once the computational and basic annotations are in place, we can move from sequences to networks of gene regulation and cell differentiation. The second part of the book begins with chapter 4, by Gary Stormo, who describes the foundations of weight matrices and their biophysical interpretation in protein-DNA interactions. In a way, this method and its variants are for regulatory motifs what the Smith-Waterman algorithm was for coding sequence comparisons. Defining the best matrix is based on the problem of defining the best multiple alignment, given the constraints of no gaps, symmetry, and other properties describing most protein-DNA binding sites in upstream regions. Abigail McGuire and George Church, in chapter 6, show how the integration of gene regulation has to be supported by experimental studies of transcriptome analyses combined with computational motif searches. Chapter 5, by Julio Collado-Vides and colleagues, is devoted to computational studies of gene regulation in *E. coli* in which different pieces are put together, making it feasible to think of a global computational study of a complete network of transcription initiation in a cell. A pair of chapters illustrate the complexity of these issues when studying eukaryotes, as seen in the signal transduction modeling by Nikolay Kolchanov and colleagues (chapter 7), and by the Boolean network methodology and its plausible application to modeling the network of factors involved in the biology of asthma by Sui Huang (chapter 8).

In chapter 9 Edward Marcotte presents a relatively novel approach using phylogenetic profiles to define a quantitative definition of function in genomics. This is a powerful method that does not require homology among genes to identify groups of genes involved in the same function. Metabolic flux analysis as well as the comparison of pathways in different genomes is illustrated in chapter 11, by Steffen Schmidt and Thomas Dandekar. The book ends with a chapter by Masaru Tomita that describes a more ambitious modeling that integrates metabolism, regulation, translation, and membrane transport. A comprehensive *in silico* complete cell model is still in its infancy, but Tomita points to what lies ahead. Still more important is evaluating the predictive capability of all these computational modeling and simulation projects.

This book does not attempt to provide a complete account of this expanding and exciting area of research. Many other databases, algorithms, and mathematical approaches are enriching postgenomic computational research. In 1995 and 1998 we participated in the organization of two Dagstuhl seminars centered on modeling and simulation of metabolism and gene regulation. This book is the outgrowth of a summer school following the Dagstuhl seminars that we organized in Magdeburg in the summer of 1999. We acknowledge the sponsorship of the Volkswagen Foundation for these activities. We also acknowledge Alberto Santos-Zavaleta and César Bonavides-Martínez for their help in editing the book. Last but not least, we are both grateful to our families for their support during the compilation of this book.

1

Are the Eyes Homologous?

Jeremy C. Ahouse

Since the 1990s research in developmental genetics has followed the approach of borrowing pathways described in one context and testing to see if the members of a pathway or genetic regulatory circuit can be found in a new context. This approach has raised questions of how the concept of homology should be used when comparing genetic regulatory circuits. One particularly cautious response has been to claim that gene expression patterns are informative for the understanding of morphological evolution *only* when coupled with a detailed understanding of comparative anatomy and embryology.

This reflects the concern that recruitment can lead to a situation where orthologous genes are expressed in novel contexts during development, thus suggesting that these similarities in gene expression patterns were not derived from a common ancestor with the structure of interest. Defining homology as a property of structures, genetic networks, or genes, rather than viewing homology as a particular way to explain observed similarities, is confusing. Specifying the similarities first and then entertaining hypotheses to explain them (including appealing to common ancestry, i.e., homology) allows us to dispense with tortured discussions of levels of biological organization at which the concept of homology may be applied.

Other chapters in this book address specific questions of gene regulation and metabolism without explicit mention of the connection between networks and the phenotype. One of the challenges, computationally, in understanding gene regulation is finding, capturing, and leveraging the information in better-studied networks. It is standard practice to apply conclusions from well-studied proteins to similar, but less well-understood, proteins. This is done when annotating for function and even when trying to predict structure (see the cautions in chapter 2 in this volume). This practice of borrowing annotations and setting expectations relies on tacit assumptions about the transitive nature of these attributes once homology has been established. It is my goal in this essay to clarify what hypotheses of homology actually are in the context of borrowing network and gene regulatory information from one (well-described) regulatory circuit to another (less well-understood).

To make the case for homology of regulatory circuits, and using what is known in one context and applying it to another, we will have to examine homology and the emergence of phenotype from regulatory circuits. This is the current challenge in computational biology. As genomes are sequenced, there comes the realization that interpreting the genome sequence is not straightforward. Coding regions are interspersed with noncoding regions, and an individual locus may give rise to multiple gene products. This has stimulated experimental approaches to identify the full spectrum of messenger RNAs (the transcriptome) and their corresponding protein products (the proteome) (RIKEN, 2001). If we now ask about the many modifications of proteins, and the numerous interactions and the detailed biophysics of protein-protein, protein-DNA, protein-RNA, and protein-lipid interactions (see chapter 9 in this volume), we quickly see why sequence-based computational biology hits a snag.

Part of the enthusiasm for moving to descriptions at the network level is the hope (or intuition) that there will be regularities that allow us to offer useful descriptions without losing the emergent biological narrative in a fog of biophysical details. In addition, the increasing availability of transcription profiles and the need to interpret them has encouraged researchers to use known regulatory networks to establish expectations against which profiling experiments can be statistically compared. I will offer an operational definition of homology, watch it at work in a current example of gene regulation (eye development), and endorse hypotheses of gene regulatory homology that push experimental work and set expectations for establishing statistical significance.

HOMOLOGY

Since evolution was championed in the mid-1800s, it has been possible to define homologies as similarities due to shared ancestry (Lankester,

Jeremy Ahouse

1870; Donoghue, 1992; Patterson, 1987; Patterson, 1988). To understand the use of this concept when thinking about developmental regulatory circuits or pathways, it is worth reflecting on the use of the term "homology." There is general agreement that attributions of homology are shorthand for the claim that particular similarities are best explained by common ancestry (Abouheif et al., 1997; Bolker and Raff, 1997; de Beer, 1971; Hall, 1995; Roth, 1984; Roth, 1988; Wagner, 1989a; Wagner, 1989b). There is still some confusion that flows from conflating "homology as an explanation for similarity" (as hypothesis) with treating homology as if it were a (discernible) property of individual things.

As more and more developmental pathway information becomes available, comparative work becomes of particular interest. I will try to provide the framework within which concepts of homology can be based in these cases. My goal is to reciprocally illuminate the comparison of regulatory pathways and those explanations that rest on homology. I will use examples from spatiotemporal gene expression patterns in developmental biology because these are the best studied. But I think much of the argument carries easily to gene regulatory circuits or metabolic pathways (see Burian, 1997 for tensions between developmental and genetic descriptions).

Here is an example. The eyespots on the wings of butterflies in the genera *Precis* and *Bicyclus* look very similar. In both species, eyespot foci are established in the larval stage. However, at the pupal stage things look quite different. The pattern of engrailed expression correlates with the development of eyespot rings. *Engrailed* is a transcription factor that is also involved in establishing body segments by activating the secreted protein *hedgehog*. In *Precis*, engrailed expression extends out to the second ring by 24 hours after pupation and then collapses to the center of the ring by 48–72 hours. In *Bicyclus*, it is expressed at the third ring but not in the second. Whereas both butterflies may use the same mechanism to place eyespots, the ways in which they specify the developing rings of the eyespot appear to be different, though the adult pattern appears similar again (Keys et al., 1999). Given the profligate reuse of transcription factors in development, we have a real challenge in applying notions of homology and in borrowing annotations from one situation to the next.

Reactions to complicated (i.e., actual) examples include the claim that homology at one level does not require homology at another, or that

Are the Eyes Homologous?

homology means nothing more than shared expression patterns of important regulatory genes during development, or that any assignment of homology must specify a level in order to be meaningful. Although homology may apply to (developmental) mechanisms per se ("process homology"), rather than to their structural end products, there is tension in the possibility that homology at one level of organization may not imply homology at another. For example, nonhomologous wings are said to have evolved from homologous forelimbs. Pterosaurs, bats, and birds share the underlying pattern of homologous forelimb bones of their tetrapod ancestor, but their wings have evolved independently. The problem is that because there is no clear way to assign levels unambiguously, one may conclude, unnecessarily, that gene expression patterns should not be used as a primary criterion of homology.

In addition to rejecting hypotheses of homology using gene expression patterns because they may disagree with each other at varying levels of organization, some critics cite specific errors that have come from using expression patterns (Abouheif et al., 1997; Bolker and Raff, 1997). These include the failure to distinguish between orthology and paralogy,[1] the confusion of analogy (convergence) and homology (noting that gene-swapping experiments do not resolve this question), the failure to notice that orthologous genes can be recruited and expressed in structures whose similarities may not be due to common ancestry. So, for example, the *distal-less* gene (the transcription factor that is the first genetic signal for limb formation to occur in the developing zygote) may be homologous in different animals, but its cis regulation may be convergent in different lineages, so that finding distal-less expression in different outgrowths does not, by itself, warrant the claim that the resultant limbs are homologous.

These concerns all seem reasonable, and might chill our enthusiasm for recognizing and borrowing knowledge gleaned from developmental regulatory circuits in different contexts. Must any hypothesis of morphological homology based on gene expression include, at a minimum, a robust phylogeny, a reconstructed evolutionary history of the gene, extensive taxonomic sampling, and a detailed understanding of comparative anatomy and embryology? Or are these requirements unnecessarily cumbersome? To untangle these issues I will return to a definition of homology.

HOMOLOGY: A DEFINITION

The use of the term "homology" implies that a given similarity is a result of common ancestry. This definition has a critical requirement: similarity comes first. There are many cases in which the similarity is cryptic, but this should not fool us into thinking that we are explaining something other than the similarity.

There are some instructive examples of structures that are not at first glance similar, but are more obviously so once the hypothesis of common ancestry is considered seriously, as in studies of insect wing evolution (Kukalova-Peck, 1983) and wing venation patterns (Kukalova-Peck, 1985). But we generally begin with the perception of similarity and then explain the similarities by appealing to a short list of possibilities. Biologists usually consider similarity to be the result of shared ancestry (homology), chance, convergence (homoplasy), or parallelism (including repeated co-optation of the same regulatory genes), or an intricate mix of these. Explanations that posit horizontal transfer are still appealing to homology to explain similarity, even though they relax the requirement for a unbroken shared lineage.

We should not appeal to homology to explain dissimilarity. And, importantly, it is not at all clear what the claim that dissimilar objects are "nonhomologous" would mean. Homology as I have defined it is coherent only when we begin with similarity. Nonhomologous similarity does make sense, however. Claiming that similarity is not due to shared ancestry sends us to the other possibilities (convergence, chance, and biomechanical constraint).

There are other uses of "homology" that we will set aside. There is the unfortunate use of the word to refer to the degree of DNA sequence identity or similarity (e.g., 30% homology). This use does not make particular claims about the origin or process that gives rise to the similarity.

Then there is the interesting phenomenon of serial homology, as in the forelimbs and hind limbs of quadrupeds, the repeated segments of a millipede, or the petals of a flower. A similar situation arises in developmental genetic terms when, for example, the expression of *apterous* in dorsal cells and *engrailed* in posterior cells in both wing and haltere discs has been taken as evidence that these two appendages are built on a "homologous groundplan" (Akam, 1998). Serial homology

Are the Eyes Homologous?

does not imply the existence of a common ancestor with just one segment, limb, or other structure; rather, it gives us insight into how a structure develops. Sometimes paralogy is assumed to be "serial homology" at the level of genes. However, paralogy of open reading frames does imply a common ancestor with just one copy.

HOMOLOGY AS HYPOTHESIS

As biologists, when we give ourselves the task to explain similarity, we have a limited list of options:

1. Mistaken perception: the similarity is solely in the eye of the beholder (flightlessness, an outgrowth, the coelom)

2. Shared ancestor had the anatomical structure, gene, regulatory network, behavior, temporal and spatial protein distribution, or other component (homology or horizontal transfer, developmental constraints)

3. Convergence, parallelism (adaptation)

4. Chance (drift, contingency, historical constraints)

5. Physical principles (biomechanics).

These options are not mutually exclusive. The claim that the perception of similarity itself is illusory is an epistemological question (and not unique to biologists), so I will put it aside. Physical constraints have been in vogue as an explanation of similarity periodically since the work of D'Arcy Thompson. Contemporary practitioners who focus on biomechanics (e.g., Mimi Koehl and Steven Vogel) are part of this tradition, as are the recent wave of neostructuralists (Webster and Goodwin, 1996; Depew and Weber, 1996). The clearest examples of this kind of similarity are in chemistry (ice crystals look similar due to the physical processes involved, not shared ancestor relationship between individual water molecules).

Physical and chemical constraints do not play a large part in most biologists' explanations, so explanations involve appeals to the other three. Much of the discussion of homology as structural, or dependent on the relative position of surrounding parts or on the percent of identical bases or amino acids comes down to questions of the relative merits of attributing overall similarity to common ancestors, not arguments about the definition of homology.

The job of explaining similarities is one of partitioning credit. Take two gene sequences that can be aligned. There will be certain positions where the residues are shared (i.e., the same). As we move along the alignment, we can imagine that some of the shared residues reflect a shared ancestor, whereas others have mutated since the common ancestor and have secondarily returned to the same residue thanks to either drift (there are only four bases possible) or to convergence (the protein works better if a particular residue is coded for at a particular position). Clearly the observation of the similarity depends strongly on the alignment (already an important hypothesis that privileges the idea that shared residues are due to homology). It should be clear that understanding what percent of the identities are due to homology, chance, and convergence may be difficult, but it is at least formally possible. Many biologists take identical residues to indicate common ancestry in combination with stabilizing selection.

Sequence comparison allows us to partition credit, at least in principle. Doing the same thing when we are discussing morphology or gene regulatory circuits is more difficult. This is both because it is much harder to atomize the trait unambiguously and because the explanations are deeply intertwined. This difficulty does not have to block inquiry.

Focusing on convergence is the traditional way to gain insight into the selectionist forces at work. Lineages are assumed to be independent trials in a natural experiment, so convergence suggests similar selection pressures (Losos et al., 1998). Alternatively, attention to the underlying homologies[2] offers insight into possible origins, and relationships among and constraints on the evolution of forms in the taxa under consideration (see Amundson, 1998 for a discussion of the structuralist tradition). Devotion to chance events has been used to good effect in both understanding the distribution and abundance of lineages and in inferring times of divergence by using background mutation rates of DNA sequences. The importance of contingent events in the history of life is well described by Gould's review of the Burgess shale fossils and his discussion of which lineages got to participate in the Cambrian explosion (Gould, 1990). These three accounts are not mutually exclusive; rather, they are the strands from which evolutionary explanations are braided.[3]

Can gene circuits and spatial and temporal expression patterns be perceived as similar? Certainly. Are they candidates for hypotheses of

Are the Eyes Homologous?

homology? I would say, absolutely yes! Now the question of diagnosis is open and difficult—but the appeals to homology, chance, and convergence as parts of an explanation are *not* especially problematic for developmental genetics (see also Gilbert et al., 1996; Gilbert and Bolker, 2001). Due to changes in developmental timing, it is often a real challenge to identify the equivalent developmental stages across lineages. Correlating equivalent developmental stages in different organisms is much like testing multiple alignment hypotheses in sequence-based comparison, though the criteria for identity are less obvious. However, if we are comparing which regulatory elements are upstream or downstream in a circuit, we can anchor our particular questions to the circuit under consideration, even before we have full resolution of the stage problem.

Can regulatory genes be homologous if the structures they produce are not? Again, I would answer this with an enthusiastic yes. I suspect that what is usually meant by "not homologous" is that the structures produced are not similar (or the part of the structures we are trying to explain are not the similarities). I find it less likely, but formally possible, that someone could convince us that the similarities of the structures are best explained by an appeal to convergence or chance or physical constraint even if the regulatory genes' similarities were best explained by their sharing a common ancestor (i.e., they are homologous). Are tissues homologous if similarity is cryptic and apparent only at level of genes? We are constantly increasing the number of ways that we can probe and understand a tissue. As should be clear by now, I would prefer to reserve assertions of homology for the actual similarities (the noncryptic gene similarities).

THE EVOLUTION OF THE EYE

The evolution of the eye stood for years as a paradigmatic example of independent evolutionary paths fulfilling the same need. Vertebrates and mollusks have single-lens eyes (though the photoreceptive cells under the lens have opposite orientation), whereas insects have compound eyes. These differences had been taken to imply that the eye evolved (independently) numerous times. We now know that the large morphological differences share a common developmental pathway of elements for optic morphogenesis. The evidence for commonality in these developmental pathways comes from looking at similar proteins in mammals and flies (the *Pax* proteins) (Gehring, 1999). A particular protein, called *eyeless* for its mutant phenotype in fruit flies, was shown to produce eye structures on wings and legs of flies when ectopically expressed in those locations. It seems reasonable to conclude that it must be near the top of the developmental hierarchy for eye development.

A mutation in a similar protein in mammals (Pax6, the eyeless homologue, based on sequence and motif similarities) results in abnormal formations of the eye. The mouse protein, when expressed in unusual locations in the fly, also results in production of ectopic fly eyes. Whether Pax6 recruits native *eyeless*, which then auto-upregulates more eyeless, or does the job itself is not known. But in either case, these two proteins have very similar functions. This finding also suggests that either (a) the common ancestor of flies and mice also had working eyes whose development used this protein (i.e., the common ancestor of Pax6 and *eyeless*) or (b) whatever this protein was doing in the common ancestor, it facilitated the evolution of eyes in other lineages (a Pax6-like protein is found in squid and octopus, too).

So are the eyes homologous? If we begin with similarities, we can avoid a fruitless argument. The differences between compound fly eyes and single-lens vertebrate eyes cannot support a hypothesis of homology because they are differences. This allows us to focus on the similarities; bilateral symmetry, positioning on the head, the expression patterns of regulatory genes, the pathway itself (*eyeless, twin of eyeless, sine oculis, eyes absent, dachshund* ...). All of these similarities do seem to be homologous; or, more carefully, we would credit those similarities to shared ancestry.

It is relevant to point out that work on the regulation of chick muscle development has shown that homologues of genes involved in mouse eye development (*Dach2, Eya2* and *Six1*) are involved in vertebrate somite (muscle) development (Heanue et al., 1999). Again by focusing on the similarities, in this case the regulatory feedback loops, we might appeal to homology while simultaneously avoiding the question of whether eyes are homologous to the segmentally organized mesodermal structures that are the embryonic precursors of skeletal muscle.

Do we need a new word for homologous gene circuits (e.g., true homology, deep homology, homoiology), or should we talk about homology at different levels? I have been arguing that attribution of similarity to historical relatedness is an appeal to homology, *whenever* it is made. The additional adjectives ("true" or "deep") do not add much.

Contingency, homology, selection (functional convergence), and physical constraints are constitutive parts of any explanation for a trait, whether it is a gene sequence, a gene expression pattern, or an adult tissue.

METHOD

While similarity surely results from a mix of explanations, a methodological preference for homology can still be defended. Looking for and highlighting homology when discussing developmental regulation serves us by generating hypotheses that inspire tests in ways that contingency and convergence do not. This does not mean that the hypothesis of homology will be supported by those tests, but we know what to do next in the laboratory.

I would like to contrast the kinds of hypotheses that are generated when we focus on differences attributed to selection rather than on similarities attributed to homology. C. J. Lowe and G. A. Wray studied several homeobox genes and concluded that they were recruited into new roles: "Each of these cases [*orthodenticle*, *distal-less*, *engrailed* expression in brittle stars, sea urchins, and sea stars] represents recruitment (co-option) of a homeobox gene to a new developmental role.... Role recruitment implies that the downstream targets are different from those in other phyla." This assessment—that if the genes were recruited into new roles, their downstream targets would be different—presents a significant experimental challenge. Where to go next? What if, instead, Lowe and Wray had asserted that the upstream and downstream factors were what had been found previously in other organisms? They would then have known which genes (and expression patterns) to hunt for. This suggests that it may be methodologically useful to hypothesize homologies, especially when looking at pathways and developmental circuits, since previously characterized networks provide a list of candidates that might be involved in the new situation.

Most evolutionists recognize that explaining every feature of an organism as an adaptation can become mere storytelling. This is why nonhomologous similarities are of special interest (i.e., distinct clades that share the feature of interest). With multiple clades, if we have ruled out homology, chance, and physical constraint, we can then look to commonalities in the respective environments to suggest that there may have been similar selection regimes. Dispensing with the comparative step can result in an uncritical adaptationism that explains (by an appeal to natural selection) the existence of a trait that is unique or novel in our lineage of interest. Without multiple lineages for comparison (focusing just on the autapomorphy) we are free to assert that the population faced whatever challenges could select for the structures under consideration.

These selectionist accounts are too difficult to challenge and can be produced at will. Flying, for example, has arisen numerous times from flightless ancestors. Should every structure that makes flight possible be treated as a complete novelty in each lineage? Because of the possibilities of finding developmental and structural homologies, there are certain parts of the explanation of flight in these lineages that will be better examined by restricting our inquiry to the three vertebrate clades that had flight (pterosaurs, birds, and bats) as distinct from the flying insects. It should be clear that comparative work is critical, and fortunately the sequencing projects and advances in transcript and protein profiling make comparative work ever easier. And the information that can be gleaned from comparative work (borrowing annotations and candidates justified by hypotheses of homology) should motivate ever more comparative studies.

From a methodological standpoint, then, identifying homologies has salutary effects. First, it demands an actual comparison. Second, in comparing across clades we can easily generate hypotheses. If our trait of interest stands in particular relations to other features in one organism—a given regulatory gene, for example—we can hypothesize that it will also do so in another. We still may not find the targets, but hypotheses of homology can tell us what to test initially.

As we move from the initial wave of genome sequencing to the wonderfully more complicated problems of understanding what proteins do, how they interact, and how they are regulated, we will need principled ways to interpret profiling information, generate network hypotheses, and annotate myriad functions. In that project, homology plays a useful role both in giving a methodological starting point for generating candidate interactions and in reminding us that inference from similarity is difficult. The use of comparative developmental genetics to generate hypotheses of homology should be embraced. Expression patterns and regulatory networks are legitimate foci for hypotheses of homology, because they help us understand the origin and evolution of structure. Finally, attributions of homology should be sought, solely on methodological grounds, because they offer us specific testable hypotheses.

ACKNOWLEDGMENTS

I would like to acknowledge pivotal conversations with Georg Halder, John True, and Jen Grenier during my postdoctoral work with Sean Carroll in the Laboratory of Molecular Biology, Howard Hughes Medical Institute, Madison, Wisconsin, and very useful comments from Kevin Padian at the Museum of Paleontology, UC Berkeley, and Scott Gilbert at Swarthmore College.

NOTES

1. The paralogy and orthology distinction was introduced to distinguish two kinds of homology in proteins (Fitch, 1970). Paralogy is meant to cover those situations when a gene duplication allows related proteins to evolve independently within the same lineage. Orthologues are found in different individuals, and paralogues can be found in the same individual (reviewed in Patterson, 1987).

2. "The importance of the science of Homology rests in its giving us the key-note of the possible amount of difference in plan within any group; it allows us to class under proper heads the most diversified organs; it shows us gradations which would otherwise have been overlooked, and thus aids us in our classification; it explains many monstrosities; it leads to the detection of obscure and hidden parts, or mere vestiges of parts, and shows us the meaning of rudiments. Besides these practical uses, to the naturalist who believes in the gradual modification of organic beings, the science of Homology clears away the mist from such terms as the scheme of nature, ideal types, archetypal patterns or ideas, &c.; for these terms come to express real facts.

The naturalist, thus guided, sees that all homological parts or organs, however much diversified, are modifications of one and the same ancestral organ; in tracing existing gradations he gains a clue in tracing, as far as that is possible, the probable course of modification during a long line of generations. He may feel assured that, whether he follows embryological development, or searches for the merest rudiments, or traces gradations between the most different beings, he is pursuing the same object by different routes, and is tending towards the knowledge of the actual progenitor of the group, as it once grew and lived. Thus the subject of Homology gains largely in interest" Charles Darwin, *On the Various Contrivances by Which British and Foreign Orchids Are Fertilised by Insects*, 2nd ed. (London: John Murray, 1877), pp. 233–234.

3. This insistence on a pluralistic account (including homology, selection, and chance) is not meant to defend claims of percent homologue. A particular similarity either is or is not homologous. The use of "homology" with respect to gene sequences to indicate percent similarity should be avoided. I am only making the uncontroversial claim that any comparison of particular traits in toto will be require an appeal to homology, convergence, and chance.

REFERENCES

Abouheif, E., Akam, M., Dickinson, W. J., Holland, P. W. H., Meyer, A., Patel, N. H., Raff, R. A., Roth, V. L., and Wray, G. A. (1997). Homology and developmental genes. *TIG* 13: 432–433.

Akam, M. (1998). Hox genes, homeosis and the evolution of segment identity: No need for hopeless monsters. *Int. J. Dev. Biol.* 42: 445–451.

Amundson, Ron (1998). Typology reconsidered: Two doctrines on the history of evolutionary biology. *Biol. Philos.* 13(2): 153–177.

Bolker, J. A., and Raff, R. A. (1997). Developmental genetics and traditional homology. *BioEssays* 18: 489–494.

Burian, R. M. (1997). On conflicts between genetic and developmental viewpoints—and their attempted resolution in molecular biology. In M. L. Dalla Chiara (ed.), *Structures and Norms in Science*. Dordrecht, The Netherlands: Kluwer Academic Publishers, pp. 243–264.

de Beer, S. G. (1971). *Homology, an Unsolved Problem*. London: Oxford University Press.

Depew, D. J., and Weber, B. (1996). *Darwinism Evolving: Systems Dynamics and the Genealogy of Natural Selection*. Cambridge, Mass.: MIT Press.

Donoghue, M. J. (1992). Homology. In E. F. Keller and E. A. Lloyd (eds.), *Keywords in Evolutionary Biology*, Cambridge, Mass.: Harvard University Press, pp. 170–179.

Fitch, W. M. (1970). Distinguishing homologous from analogous proteins. *Syst. Zool.* 19: 99–113.

Gehring, W. J. (1999). Pax 6 mastering eye morphogenesis and eye evolution. *TIG* 15(9): 371–377.

Gilbert, S. F., and Bolker, J. A. (2001). Homologies of process: Modular elements of embryonic construction. In G. Wagner (ed.), *The Character Concept in Evolutionary Biology*. New Haven, Conn.: Yale University Press, pp. 435–454.

Gilbert, S. F., Opitz, J., and Raff, R. A. (1996). Resynthesizing evolutionary and developmental biology. *Dev. Biol.* 173: 357–372.

Gould, Stephen Jay. (1990). *Wonderful Life: The Burgess Shale and the Nature of History*. New York: W. W. Norton.

Hall, B. K. (1995). Homology and embryonic development. *Evol. Biol.* 28: 1–37.

Heanue, T. A., Reshef, R., Davis, R. J., Mardon, G., Oliver, G., Tomarev, S., Lassar, A. B., and Tabin, C. J. (1999). Synergistic regulation of vertebrate muscle development by

Dach2, Eya2, and Six1, homologs of genes required for *Drosophila* eye formation. *Genes and Dev.* 13: 3231–3243.

Keys, D. N., Lewis, D. L., Selegue, J. E., and Carroll, S. B. (1999). Recruitment of a hedgehog regulatory circuit in butterfly eyespot evolution. *Science* 283: 532–534.

Kukalova-Peck, J. (1983). Origin of the insect wing and wing articulation from the arthropodan leg. *Can. J. Zool.* 61: 1618–1669.

Kukalova-Peck, J. (1985). Ephemeroid wing venation based upon new gigantic Carboniferous mayflies and basic morphology, phylogeny, and metapmorphosis of pterygote insects (Insecta, Ephemerida). *Can. J. Zool.* 63: 933–955.

Lankester, E. R. (1870). On the use of the term homology in modern zoology. *Annu. Mag. Nat. Hist.* ser. 4, 6: 34–43.

Losos, J. B., et al. (1998). Contingency and determinism in replicated adaptive radiations of island lizards. *Science* 279 (March 27): 2115–2118.

Lowe, C. J., and Wray, G. A. (1997). Radical alterations in the roles of homeobox genes during echinoderm evolution. *Nature* 389: 718–721.

Patterson, C. (1982). Morphological characters and homology. In K. A. Joysey and A. E. Friday (eds.), *Problems of Phylogenetic Reconstruction*. London: Academic Press, pp. 21–74.

Patterson, C. (1987). Introduction. In C. Patterson (ed.), *Molecules and Morphology in Evolution: Conflict or Compromise?*. Cambridge: Cambridge University Press.

Patterson, C. (1988). Homology in classical and molecular biology. *Mol. Biol. Evol.* 5: 603–625.

RIKEN Genome Exploration Research Group Phase II Team and the FANTOM Consortium. (2001). Functional annotation of a full-length mouse cDNA collection. *Nature* 409: 685–690.

Roth, V. L. (1984). On homology. *Biol. J. Linnaean Soc.* 22: 13–29.

Roth, V. L. (1988). The biological basis of homology. In C. J. Humphries (ed.), *Ontogeny and Systematics*. New York: Columbia University Press, pp. 1–26.

Wagner, G. P. (1986). The systems approach: An interface between development and population genetic aspects of evolution. In D. M. Raup and D. Jablonski (eds.), *Patterns and Processes in the History of Life*. Berlin: Springer-Verlag, pp. 149–165.

Wagner, G. P. (1989a). The biological homology concept. *Annu. Rev. Ecol. Systemat.* 20: 51–69.

Wagner, G. P. (1989b). The origin of morphological character and the biological basis of homology. *Evolution* 43: 1157–1171.

Webster, G., and Goodwin, B. (1996). *Form and Transformation: Generative and Relational Principles in Biology*. Cambridge: Cambridge University Press.

SUGGESTED READING

Carroll, Sean B., Grenier, Jennifer K., and Weatherbee, Scott D. (2001). *From DNA to Diversity: Molecular Genetics and the Evolution of Animal Design*. Oxford: Blackwell Science.

Davidson, Eric H. (2001). *Genomic Regulatory Systems: Development and Evolution*. London: Academic Press.

Gehring, Walter J., and Ruddle, Frank. (1998). *Master Control Genes in Development and Evolution: The Homeobox Story*. New Haven, Conn.: Yale University Press. (Terry Lectures).

Gerhart, John, and Kirschner, Marc W. (1997). *Cells, Embryos, and Evolution: Toward a Cellular and Developmental Understanding of Phenotypic Variation and Evolutionary Adaptability*. Oxford: Blackwell Science.

Gilbert, Scott F. (2000). *Developmental Biology*. 6th ed. Sunderland, Mass.: Sinauer Associates.

Hall, Brian K. (1998). *Evolutionary Developmental Biology*. London: Chapman & Hall.

Hall, Brian K. (2000). *Homology: The Hierarchical Basis of Comparative Biology*. London: Academic Press.

Lawrence, Peter A. (1992). *The Making of a Fly: The Genetics of Animal Design*. Oxford: Blackwell Science.

Owen, Richard, and Sloan, Phillip Reid (eds.). (1992). *The Hunterian Lectures in Comparative Anatomy, May–June, 1837*. Chicago: University of Chicago Press.

Raff, Rudolf A. (1996). *The Shape of Life: Genes, Development, and the Evolution of Animal Form*. Chicago: University of Chicago Press.

Sober, E. (1988). *Reconstructing the Past*. Cambridge, Mass.: MIT Press.

Wiley, E. O., Siegel-Causey, D., Brooks, D. R., and Funk, V. A. (1991). *The Compleat Cladist: A Primer of Phylogenetic Procedures*. Lawrence: University of Kansas, Museum of Natural History.

URLs FOR RELEVANT SITES

Flybase: The Interactive Fly describes fly proteins and their actions and interactions. `http://flybase.harvard.edu:7081/allied-data/lk/interactive-fly/aimain/1aahome.htm`

The Hennig Society. If your work with homologies brings you to constructing trees, you will want to explore cladistics. `http://www.cladistics.org/education.html`

Kyoto Encyclopedia of Genes and Genomes (KE66) is an effort to capture molecular and cellular biology in terms of the information pathways that consist of interacting proteins. `http://www.genome.ad.jp/kegg/`

Virtual Embryo is a collection of developmental biology tutorials and links. `http://www.ucalgary.ca/UofC/eduweb/virtualembryo`

I Information and Knowledge Representation

2 Automation of Protein Sequence Characterization and Its Application in Whole Proteome Analysis

Rolf Apweiler, Margaret Biswas, Wolfgang Fleischmann, Evgenia V. Kriventseva, and Nicola Mulder

The first complete genome sequence of an organism, the five-kilobase sequence of the bacterial virus phi-X174, was achieved by Fred Sanger and coworkers in Cambridge (Sanger et al., 1978). Only more recently, however, has the technology developed to a stage where the sequencing of the complete genome of a living organism can be contemplated as a practical and routine possibility. A major breakthrough was the sequencing of the first complete eukaryote chromosome, chromosome III of *Saccharomyces cerevisiae*, in 1992 by a European Union-funded consortium (Oliver et al., 1992). In 1995 the TIGR group published the first complete sequence of a bacterial genome, that of *Haemophilus influenzae* (Fleischmann et al., 1995).

Since those dramatic events the complete sequences of more than 40 bacterial genomes have been published and at least 70 more are known to be nearing completion. The sequencing of five eukaryotic genome sequences—those of *Saccharomyces cerevisiae* (Goffeau et al., 1997), the nematode *Caenorhabditis elegans* (The C. elegans Consortium, 1998), the fruit fly *Drosophila melanogaster* (Adams et al., 2000), the plant *Arabidopsis thaliana* (The Arabidopsis Initiative, 2000), and the alga *Guillardia theta* (Douglas and Penny, 1999) has been achieved and the sequences of other model eukaryotes are nearing completion. Large-scale sequencing of the genome of the laboratory mouse is well under way in the United States, Japan, and Europe. The sequences of several important protozoan parasites are close to being finished. In addition, the complete genomes of many mitochondria and plastids have been determined. The "Holy Grail" of large-scale sequencing is, however, the determination of the sequence of the human genome, estimated at around 3 billion base pairs. The completion of the "first draft" of this sequence was announced on 26 June 2000 by an international consortium of public laboratories.

Various proteomics and large-scale functional characterization projects in Europe, Japan and the United States complement the large-scale nucleotide sequencing efforts. These projects have all produced large amounts of sequence data lacking experimental determination of the biological function. To cope with such large data volumes and to provide meaningful information, new approaches to characterize and annotate the biological data in a faster and more effective way are required. One promising but still error-prone approach is automatic functional analysis, which is generated with limited human interaction.

AUTOMATIC ANNOTATION

The Pitfalls of Automatic Functional Analysis

Several solutions of automatic functional characterization of unknown proteins are based on high-level sequence similarity searches against known proteins. Other methods collect the results of different prediction tools in a simple (`http://pedant.gsf.de/`; Frishman and Mewes, 1997) or a more elaborate (`http://jura.ebi.ac.uk:8765/ext-genequiz/`; Tamames et al., 1998; Hoersch et al., 2000) manner. However, some of the currently used approaches have several drawbacks, including the following:

· Since many proteins are multifunctional, the assignment of a single function, which is still common in genome projects, results in the loss of information and outright errors.

· Since the best hit in pairwise sequence similarity searches is frequently a hypothetical protein, a poorly annotated protein, or simply a protein that has a different function, the propagation of wrong annotation is widespread.

· There is no coverage of position-specific annotation, such as active sites.

· The annotation is not constantly updated, and thus is quickly outdated.

It is also important to emphasize that a single sentence describing some predicted properties of an unknown protein should not be regarded as annotation. It may be regarded as an attempt to characterize a protein, but not as an attempt to annotate the protein. Annotation means the addition to a protein sequence of as much reliable and up-to-date information as possible describing properties such as function(s) of the protein, domains and sites, catalytic activity, cofactors, regulation, induction, subcellular location, quaternary structure, diseases associated with deficiencies in the protein, the tissue specificity of a protein, developmental stages in which the protein is expressed, pathways and processes in which the protein may be involved, similarities to other proteins, and so on.

The Annotation Concept of SWISS-PROT and TrEMBL

The SWISS-PROT protein sequence database (Bairoch and Apweiler, 2000) strives to provide extensive annotation as defined above. The increased data flow from genome projects to the protein sequence databases, however, challenges this time- and labor-intensive method of database annotation. Maintaining the high quality of annotation in SWISS-PROT requires the careful and detailed annotation of every entry with information retrieved from the scientific literature and from rigorous sequence analysis. This is the rate-limiting step in the production of SWISS-PROT. It is of paramount importance to maintain the high editorial standards of SWISS-PROT because the exploitation of the sequence avalanche is heavily dependent on reliable data sources as the basis for automatic large-scale functional characterization and annotation by comparative analysis. This, then, sets a limit on how much the SWISS-PROT annotation procedures can be accelerated. Recognizing that it is also vital to make new sequences available as quickly as possible, in 1996 the European Bioinformatics Institute (EBI) introduced TrEMBL (*Tr*anslation of *EMBL* nucleotide sequence database). TrEMBL consists of computer-annotated entries derived from the translation of all coding sequences (CDS) in the EMBL database, except for CDS already included in SWISS-PROT.

To enhance the annotation of uncharacterized protein sequences in TrEMBL, the SWISS-PROT/TrEMBL group at the EBI developed a novel method for automatic and reliable functional annotation (Fleischmann et al., 1999). This method selects proteins in the SWISS-PROT protein sequence database that belong to the same group of proteins as a given unannotated protein, extracts the annotation shared by all functionally characterized proteins of this group, and assigns this common annotation to the unannotated protein.

Automatic Annotation of TrEMBL

To implement this methodology for the automated large-scale functional annotation of proteins, three major components are required. First, a reference database must serve as the source of annotation. SWISS-PROT makes an excellent reference database due to its highly reliable, well-annotated, and standardized information. Second, a highly reliable, diagnostic protein family signature database must provide the means to assign proteins to groups. Initially, PROSITE (Hofmann et al., 1999) was used, and in future, InterPro, described below, will be used. The third component needed for the implementation of the automated large-scale functional annotation methodology is a database (RuleBase) that stores and manages the annotation rules, their sources, and their usage.

The Reference Database   The basis for the automatic annotation of TrEMBL is the functional information in the SWISS-PROT protein sequence database. Many other annotation approaches try to predict functions by comparative analysis with SWISS-PROT and other protein databases like TrEMBL and Genpept. There are three main reasons for using only SWISS-PROT annotation in automatic approaches.

First, SWISS-PROT is a comprehensive protein sequence database. This may seem surprising, since as of October 2000 SWISS-PROT contains only 88,000 proteins. Although these sequences represent—taking redundancy into account—less than one-third of all known protein sequences, SWISS-PROT contains around 60% of all proteins found in comprehensive protein sequence databases (like SWISS-PROT+ TrEMBL [SPTR] or protein entries in Entrez) with annotation of at least basic experimentally derived functional characterization. This percentage was estimated from the number of papers (70,000) cited in SWISS-PROT records compared with the number of papers in all SPTR or Entrez protein entries (110,000) together. The calculation was based on the assumption that the proportion of papers reporting sequencing to papers reporting characterization is the same in SWISS-PROT records as in TrEMBL records or in non–SWISS-PROT Entrez protein records. However, an inspection of citations from SWISS-PROT compared with citations from TrEMBL shows that SWISS-PROT contains a higher proportion of papers representing biochemical citation than do TrEMBL papers.

This observation, together with the sequence redundancy in TrEMBL and the non–SWISS-PROT records of Entrez proteins, indicates that SWISS-PROT probably contains more than 60% of all annotated proteins with at least basic biochemical characterization. Even more striking is the fact that more than 80% of all functional annotation found in the comprehensive protein sequence database records (such as SPTR or protein entries in Entrez) is SWISS-PROT annotation. SWISS-PROT annotation is, for the most part, stored in the CC (Comment), FT (Feature Table), KW (Keyword) and DE (Description) lines. As of August 2000, there are more than 410,000 CC lines, 460,000 FT lines, and 110,000 DE lines in SWISS-PROT. This information in SWISS-PROT is abstracted from more than 70,000 literature citations reporting sequencing and/or characterization.

Another important reason is the standardization of annotation in SWISS-PROT. This unique feature of SWISS-PROT allows the extraction of the "common annotation" described above. Using the standardized SWISS-PROT annotation leads eventually to the standardized annotation of TrEMBL.

The last and perhaps most important reason is the fact that SWISS-PROT distinguishes experimentally determined functions from those determined computationally.

InterPro   InterPro (Apweiler et al., 2001) is an integrated resource for protein families, domains, and functional sites, developed as an integrative layer on top of the PROSITE, PRINTS (Attwood et al., 2000), Pfam (Bateman et al., 2000), and ProDom (Corpet et al., 2000) databases. The different approaches integrated in InterPro (hidden Markov models [HMMs], profiles, fingerprints, regular expressions, etc.) have different strengths and weaknesses. The combination of the strengths of the different signature recognition methods, coupled with a statistical and biological significance test, overcomes drawbacks of the individual methods. InterPro reliably classifies proteins into families and recognizes the domain structure of multidomain proteins. The use of InterPro should facilitate increased coverage of target sequences with enhanced reliability (reduction of false positives and false negatives). InterPro can currently classify around 60% of all known protein sequences.

RuleBase RuleBase stores the common annotation extracted from a group of SWISS-PROT entries. The common annotation is linked to the conditions and to the set of proteins from which the annotation was derived. The concept of a rule is used so that every rule has one or more conditions and one or more actions associated with it. If the conditions hold for a target TrEMBL entry, then all the actions are applied to that entry (Fleischmann et al., 1999).

Implementation The actual flow of information during automatic annotation can be divided into five steps.

1. Use InterPro and additional a priori knowledge to extract the information necessary to assign proteins to groups (conditions) and store the conditions in RuleBase.

2. Group the proteins in SWISS-PROT by the stored conditions.

3. Extract from SWISS-PROT the common annotation shared by all functionally characterized proteins from each group. Store this common annotation together with its conditions in RuleBase. Every rule consists of conditions and the annotation common to all proteins of the group characterized by these conditions.

4. Group the unannotated, target TrEMBL entries by the conditions stored in RuleBase.

5. Add the common annotation to the unannotated TrEMBL entries. The predicted annotation will be flagged with evidence tags, which will allow users to recognize the predicted nature of the annotation as well as the original source of the inferred annotation.

Because the reliability of the conditions is crucial to the reliability of the methodology, measures are taken to minimize false-positive automatic annotation. The InterPro database that is used to extract conditions and to assign proteins to groups integrates different computational techniques for the recognition of signatures that are diagnostic for different protein families or domains. In addition, every rule ensures that the taxonomic classification of the unannotated protein sequences lies within the known taxonomic range of the experimentally characterized proteins.

This automatic annotation approach should overcome some limitations of some existing automatic annotation methods in the following ways:

- By using only the annotation from a reliable reference database for the predictions, the propagation of wrong annotation, one of the core problems in functional annotation, is drastically reduced (Bork and Koonin, 1998).

- By using the "common annotation" of multiple entries, the implemented methodology will produce a significantly lower number of overpredictions than methods based on the best hit of a sequence similarity search.

- Using the "common annotation" from a reliable reference database with standardized annotation and nomenclature ensures the standardized annotation of uncharacterized, target proteins by avoiding the use of wrong nomenclature and of different descriptions for the same biological fact.

- Since the method takes all potential annotation available in the reference database into account, a much higher level of annotation, including position-specific annotation such as active sites, is possible.

- The "common annotation" approach can be used not only with protein families but also with conditions aiming at a higher level in the protein family hierarchy. Only the annotation common to all members of this (for instance) superfamily will be copied over.

- Our methodology is independent of the multidomain organization of proteins. If a certain condition aims at a single domain that occurs with various other domains, it can be expected that only the annotation referring to this single domain will be found in all relevant characterized proteins. On the other hand, if the single domain always occurs with another domain, the information for the other domain will be picked up as well.

- Evidence tags will allow the automatic update of the predicted annotation if the underlying conditions or the "common annotation" in RuleBase changes.

WHOLE PROTEOME ANALYSIS

A Four-Layer Approach to Whole Proteome Analysis

It is no longer ludicrous to envisage collecting vast amounts of genomic data, although it remains a massive task. The challenge is in developing the tools and methods required to analyze the data. In the sections above, we described how the SWISS-PROT group at the EBI combines manual annotation and sequence analysis of SWISS-PROT entries with rule-based automatic annotation of TrEMBL entries to provide a comprehensive, reliable, and up-to-date protein sequence database. With existing methodology we are able to improve the annotation of approximately 25% of the incoming data. Exploiting this approach to the full will enable us to annotate approximately 40–50% of the new and existing sequence data in a reasonable way within a few years. However, tools developed by our group and by others make possible the preliminary classification and characterization of many more sequences. Capitalizing on these achievements, we developed a new four-layer strategy for protein analysis:

1. Automatic protein classification

2. Automatic protein characterization

3. Rule-based automatic annotation

4. SWISS-PROT-style manual annotation.

From level 1 to level 4 there is an increase in the manual intervention required and a decrease in both the computational power needed and the number of protein sequences affected. The rule-based automatic annotation of TrEMBL entries and the SWISS-PROT-style manual annotation (levels 3 and 4) were described above. In the following sections we will describe automatic protein classification and characterization, and their application to provide statistical and comparative analysis, as well as structural and other information, for complete proteome sequences.

Whole Proteome Analysis at EBI

The EBI proteome analysis initiative aims to provide comprehensive, easily accessible information as quickly as possible to the user community. Proteome analysis data have been produced for all the completely sequenced organisms spanning archaea, bacteria, and eukaryotes. Complete proteome sets for each organism have been assembled from the SPTR (SWISS-PROT+TrEMBL+TrEMBLnew) database (Apweiler, 2000) to be wholly nonredundant at the sequence level. These proteome data have been used in the analysis, and are easily accessible and downloadable from the proteome analysis pages (`http://www.ebi.ac.uk/proteome/`).

Automatic Protein Classification

For the automatic classification of proteins, InterPro (Apweiler et al., 2001), CluSTr, HSSP (Sander and Schneider, 1991), TMHMM (Sonnhammer et al., 1998), and SignalP (Nielsen et al., 1999) are used. SignalP is used for the prediction of signal peptides and their cleavage sites in eukaryotes and prokaryotes in order to classify secreted proteins and transmembrane proteins with signal sequences. TMHMM predicts transmembrane helices in proteins and is used for the identification and classification of transmembrane proteins. A list of nonredundant proteins from the reference proteome with HSSP (homology-derived secondary structure of proteins) links has been generated from current releases of SWISS-PROT and TrEMBL. These proteins, together with those having a corresponding PDB (Berman et al., 2000) entry, represent the proteins with structural classification.

The resources with the highest information content are InterPro and CluSTr. InterPro (`http://www.ebi.ac.uk/interpro/`) classifies 50–70% of all proteins in a proteome into distinct families. In addition, InterPro provides insights into the domain composition of the classified proteins. The proteome analysis pages (`http://www.ebi.ac.uk/proteome/`) make available InterPro-based statistical analysis that includes the following, among other information:

• General statistics—lists all InterPro entries with matches to the reference proteome. The matches per genome and the number of proteins matched for each InterPro entry are displayed.

• Top 30 entries—lists the 30 InterPro entries with the highest number of protein matches for the reference proteome.

• 15 most common domains—lists the InterPro entries with the largest number of Pfam and profile matches (defined as domains) for the reference proteome. The matches per genome and the number of proteins matched for each InterPro entry are shown.

CluSTr

There are several databases that focus on the analysis of complete protein sequences. The COG database (Clusters of Orthologous Groups of proteins) is a phylogenetic classification of proteins encoded in 21 complete genomes of bacteria, archaea, and eukaryotes (`http://`

www.ncbi.nlm.nih.gov/COG; Tatusov, 2000). ProtoMap offers a hierarchical classification of proteins in the SWISS-PROT and TrEMBL databases (http://www.protomap.cs.huji.ac.il/; Yona et al., 2000) based on analysis of all pairwise similarities among the protein sequences. The searching algorithm SYSTERS (SYSTEmatic Re-Searching) applies an iterative method for database searching to cluster sequences from a number of databases that store protein sequences (http://www.dkfz-heidelberg.de/tbi/services/cluster/systersform; Krause et al., 2000).

CluSTr (http://www.ebi.ac.uk/clustr/), the database of clusters of SWISS-PROT and TrEMBL proteins developed at EBI, will be discussed in some detail in this chapter. It offers an automatic classification of SWISS-PROT+TrEMBL (SPTR) proteins into groups of related proteins. The clustering is based on analysis of all pairwise comparisons between protein sequences. Analysis has been carried out for different levels of protein similarity, yielding a hierarchical organization of clusters.

Methodology

The clustering approach is based on two steps. First, a similarity matrix of "all-against-all" protein sequences is built. The similarity matrix is computed using the Smith-Waterman algorithm (Smith and Waterman, 1981). A Monte Carlo simulation, resulting in a Z-score (Comet et al., 1999), is used to estimate the statistical significance of similarity between potentially related proteins. Initially, a Smith-Waterman score between sequences A and B is calculated. If this score is higher than a certain threshold, sequence A is compared with N shuffled sequences of B (B^*). Sequences B^* have the same length and amino acid composition as the initial sequence B. The Z-score is calculated as

$$Z(A, B) = \frac{SW(A, B) - M}{\sigma},$$

where, $SW(A, B)$ is the initial Smith-Waterman score, M is the average Smith-Waterman score between sequence A and sequences B^*, and σ is the standard deviation. Sequence B is then compared with N shuffled sequences of A (A^*) and $Z(B, A)$ is calculated. The final Z-score is

Z-score $= \min(Z(A, B), Z(B, A))$.

The Z-score depends only on the compared sequences, not on the size and composition of the sequence database. By storing all the scores of unchanged sequences and calculating only "new-against-new" and "new-against-unchanged," the CluSTr database can be updated incrementally, avoiding time-consuming recalculations.

Second, clusters are built using a single linkage algorithm (Sneath and Sokal, 1973) for different levels of protein similarity. There are two major complications in automatic clustering procedures: different protein families have different levels of sequence similarity, and clusters of proteins with different domains get pulled together by multidomain proteins. One of the approaches to tackling these problems uses hierarchical clustering that works with clusters at different levels of sequence similarity. The LASSAP package (Glemet and Codani, 1997) has been used to calculate similarities and to build clusters.

Data Structure

Clusters for mammalian proteins, plant proteins, and the three complete eukaryotic genomes (*Caenorhabditis elegans*, *Saccharomyces cerevisiae*, and *Drosophila melanogaster*) have been built. The CluSTr data are stored in a relational database that comfortably handles large amounts of data and facilitates comprehensive data updates. Multiple users have direct access to the database via Java servlets.

The main building blocks of the schema underlying the CluSTr are Proteins, Groups, Similarities, and Clusters. The Proteins table describes SPTR entries, Groups describes protein sets for which clusters were built and the history of comparison runs, Similarities contains the pairwise scores between proteins, and Clusters represents the information about and relationships between different clusters.

Keeping the data up-to-date has been another big challenge in the design and implementation of the CluSTr database. The aim is to update the CluSTr data incrementally, in a synchronized manner with the weekly updates of SPTR. There are additional Oracle tables to facilitate this. The Protein_New table gets populated with new protein data. New, changed, and deleted proteins are checked for, using SPTR accession numbers and the circular redundancy check sum (CRC64). An algorithm to compute the CRC64 is described in the ISO-3309 standard (ISO-3309, 1993). While, in theory, two different sequences could have the same CRC64 value, the likelihood that this would happen is quite low. A list of new and changed proteins is created, and the calculation of similarities for this new set against itself and against unchanged proteins is carried out.

User Interface

The CluSTr database is available for querying and browsing at http://www.ebi.ac.uk/clustr. A query can be made using SPTR accession numbers, SWISS-PROT ID entry names, sequence annotation, key words, and taxonomic information. The result of the query is a graphical presentation of corresponding clusters at different levels of protein similarity. For example, the results for a text query of "human sodium transport" proteins are shown in figure 2.1. On the right of the

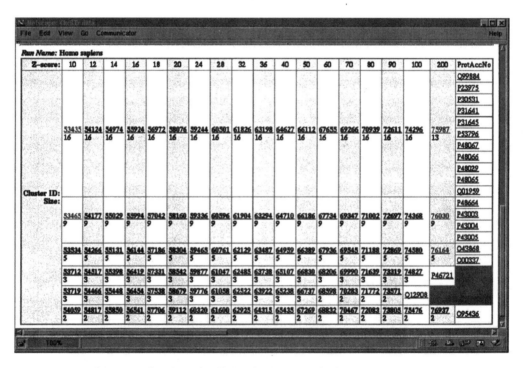

Figure 2.1 Searching the CluSTr database. Results for a query of "human sodium transport" proteins. The table contains accession numbers of proteins with the words "human" and "sodium transport" in their annotation and the corresponding clusters at different z-levels.

table are accession numbers of proteins with the words "human" and "sodium transport" in their annotation, and on the left is the cluster structure which these proteins form at different Z-levels. Bigger groups of clusters of size 16, 9, and 5 correspond to Sodium:neurotransmitter symporter family (IPR000175), Sodium:dicarboxylate symporter family (IPR001991) and Na^+ dependent nucleoside transporter (IPR002668), respectively. The next group of proteins is not well described. At the bottom of the table are the sodium bile acid symporter family (IPR002657) and sodium-dependent phosphate transport proteins.

A cluster of interest can be further investigated by clicking on its ID number. For each cluster the list of proteins, their descriptions, and the domain composition is provided (figure 2.2). Links to the Sequence Retrieval System (SRS) (Etzold et al., 1996) allow users to download the list of proteins from a cluster. The domain composition is defined using InterPro. Links to the InterPro graphical view allow users to see at a glance whether proteins from a particular cluster share common domains or functional sites (figure 2.3). For each cluster a list of secondary structure cross-references from the Homology-derived Secondary Structure of Proteins (HSSP) database (Sander and Schneider, 1991) is generated dynamically. The database also provides links to the Protein Data Bank (PDB) (Berman et al., 2000).

It has already been mentioned that the clusters are built for specific taxonomic groups. For each of the organisms that have been studied, the following information is displayed:

• General statistics—the number of clusters with two or more proteins, the total number of proteins in these clusters, the number of singletons, and the number of distinct families at different levels of protein similarity.

• List of singletons—proteins that form clusters of size 1 at the lowest studied protein similarity level.

• 30 biggest clusters—the 30 biggest protein clusters and their InterPro-based functional classification.

• Clusters without InterPro links—clusters of size 5 or more which have no matching InterPro families, domains, or functional sites.

• Clusters without HSSP links—clusters of size 5 or more for which there are no HSSP matches.

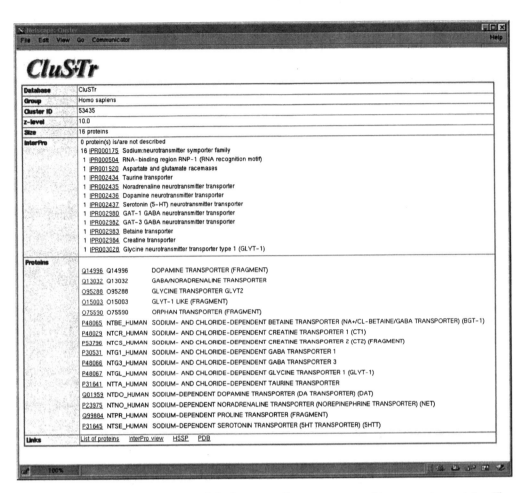

Figure 2.2 A cluster of the human sodium: neurotransmitter symporter proteins. The presentation contains general information, a list of proteins, their description, and an InterPro-based domain description of the cluster. At the bottom of the page are links to the SRS-generated list of clustered proteins, the InterPro graphical representation, and links to the HSSP and PDB databases.

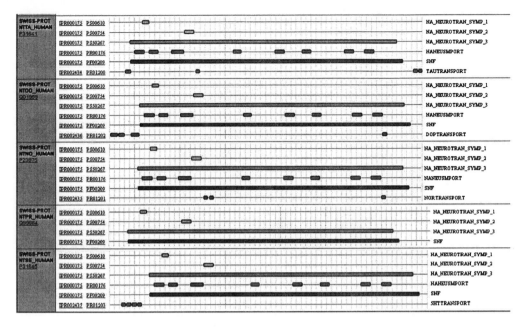

Figure 2.3 Part of the InterPro graphical view for the cluster of the human sodium: neurotransmitter symporter proteins (ID 53435) from figure 2.2).

AUTOMATIC PROTEIN CHARACTERIZATION

The Gene Ontology (GO) (Ashburner et al., 2000), created by FlyBase (The FlyBase Consortium, 1999), Saccharomyces Genome Database (SGD) (Ball et al., 2000), and Mouse Genome Database (MGD) (Blake et al., 2000) is gaining acceptance as a universal controlled vocabulary to annotate genes and gene products. GO terms are currently being assigned to proteins in SWISS-PROT and TrEMBL, and to InterPro domains and families. Before the GO mapping began, each InterPro entry was assigned a functional classification in the form of a three-letter code with the categories based on top-level GO terms. Using this basic classification, SWISS-PROT key words, and manual inspection of annotation of protein families, specific GO terms of all levels were mapped to each InterPro entry.

This mapping is in progress, and of the three organizing principles of GO (details can be found at http://www.geneontology.org/), biological processes and molecular function have been taken up first. If the number of proteins with a known or reliably predicted subcellular location becomes significant, the cellular component of GO will also be included in the classification. There are cases where InterPro entries describe nonspecific protein domains or families that cannot be assigned a specific GO term, and in addition there are cases where GO terms do not yet exist (e.g., when the domain or family is specific to prokaryotes). These InterPro entries, however, still contain the three-letter functional classification code, which may be more general and includes the categories Unknown Function and Multifunctional Proteins.

Using the classification data and selecting only the top-level terms in the GO hierarchy, a table has been created for each completed proteome that lists the GO terms and the number of proteins mapped to each term. These tables can be found through links from the proteome analysis pages for each organism. For *Drosophila melanogaster*, for example, the page is located at http://www.ebi.ac.uk/proteome/ DROME/go/function.html. A functional classification of the proteins within each proteome set has thus been generated to show the percentage of proteins involved in, for example, metabolism, transcription, and so on. This is represented graphically for three eukaryotes in figure 2.4. The functional classification and mapping to GO of InterPro families and domains is a simple method for determining whole proteome composition and provides a basis for comparative analysis. It also provides a framework for the mapping to GO of all proteins in SWISS-PROT and TrEMBL that have matches in InterPro, and for any new or previously uncharacterized protein sequences searched for in InterPro. In addition, the CluSTr database has links to InterPro, and from there to the corresponding functional classification codes and GO terms, thus making it possible to identify protein functions within clusters.

COMPARATIVE PROTEOME ANALYSIS

Some of what are likely to be the most frequently requested comparisons are available from the index page for each of the reference proteomes. Comparisons of the proteomes are based on InterPro statistics and are precomputed. The proteomes for which such comparisons are currently available are the archaea; *Pyrococcus abyssi* and *Pyrococcus horikoshii*; various groups of bacteria: *Bacillus subtilis* and *Escherichia coli*; *Chlamydia pneumoniae*, *Chlamydia trachomatis*, and *Chlamydia mur-*

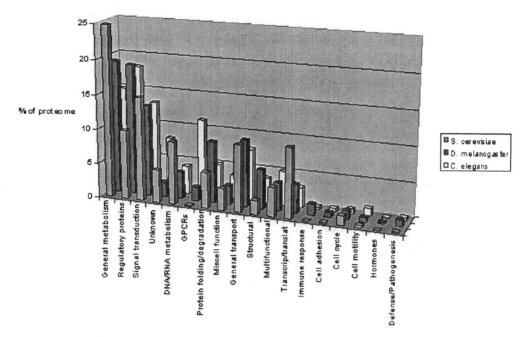

Figure 2.4 Relative representation of different protein functions in the three complete eukaryotic proteomes based on the InterPro classification system. GPCRs = the G-protein coupled receptors.

idarum; the two *Helicobacter pylori* strains (26695 and J99); and *Mycoplasma genitalium* and *Mycoplasma pneumoniae*; and the three complete eukaryotic proteomes *Caenorhabditis elegans*, *Drosophila melanogaster*, and *Saccharomyces cerevisiae*. The incomplete proteome of *Homo sapiens* has been compared with three complete eukaryotic proteomes, and the resulting data are available from the index page for *Homo sapiens* (http://www.ebi.ac.uk/proteome/HUMAN/).

Interactive InterPro-based comparisons can be made using the InterPro proteome comparisons program to select the proteomes of the organisms to be compared and the type of comparative analysis to be carried out (http://www.ebi.ac.uk/proteome/comparisons. html). Comparisons that can be made include general statistics, top 30 entries, top 200 entries, 10 biggest protein families, and 15 most common domains. An additional feature is the option to compute a list of shared InterPro entries that are common to all the selected proteomes (similar in concept to the overlapping region of a Venn diagram).

SOME OBSERVATIONS FROM THE COMPARATIVE PROTEOME ANALYSIS OF *CAENORHABDITIS ELEGANS, DROSOPHILA MELANOGASTER,* AND *SACCHAROMYCES CEREVISIAE*

A comparative analysis of the three complete eukaryotic proteomes was the first application of some of the resources described here. The InterPro analysis plus manual data inspection enabled the assignment of just over 50% of the proteins of the proteomes of *Drosophila melanogaster, Caenorhabditis elegans,* and *Saccharomyces cerevisiae* (Rubin et al., 2000). The Proteome Analysis Database (`http://www.ebi.ac.uk/proteome/`) now contains the comparative analysis of all complete proteomes. The analysis is carried out using complete nonredundant proteome sets that comprise records taken from the SWISS-PROT, TrEMBL, and TrEMBLnew databases (Apweiler, 2000) corresponding to the complete proteome. The proteome sets are wholly nonredundant at the sequence level (`http://www.ebi.ac.uk/proteome/CPhelp.html`). The average protein length and size range of full-length proteins (excluding fragments) for each of the three eukaryotic proteomes are presented in table 2.1. The average length of the proteins is similar in all three organisms, and higher than the average length of bacterial proteins (unpublished observation).

The smallest proteins in *Saccharomyces cerevisiae* are the 60S ribosomal protein L41 (P05746) and the leader peptide CPA1 (P08521), while the largest is a hypothetical 560 kDa protein (Q12019). The largest protein from *Caenorhabditis elegans* is a 1368.6 kDa uncharacterized protein (Q09165) that contains several domains and motifs, including fibronectin type III repeats, a Von Willebrand factor type A domain, and EGF-like domains. The smallest *Drosophila melanogaster* protein

Table 2.1 A comparison of the average protein length and size range of full-length proteins for each of the three eukaryotic proteomes

Proteome	Number of amino acid residues	
	Average protein length	Size range
Saccharomyces cerevisiae	476.69 ± 375.13	25 to 4910
Caenorhabditis elegans	434.76 ± 384.38	20 to 13,055
Drosophila melanogaster	486.91 ± 451.66	8 to 7182

Rolf Apweiler et al.

(Q9VRD2) is predicted to be just 8 amino acids long and is known as the CG11666 protein.

For each of the proteomes SignalP (Nielsen et al., 1999), a signal peptide prediction program, was run to find all proteins that are localized in the membrane or secreted. The transmembrane proteins were identified using the transmembrane prediction program TMHMM version 1.0 (Sonnhammer et al., 1998), and the secreted soluble proteins were classified based on the prediction of signal proteins adjusted to remove those that are membrane proteins. The percentage of the proteome found to be secreted proteins was 21.6% for *Caenorhabditis elegans*, 20.1% for *Drosophila melanogaster*, and 12.7% for *Saccharomyces cerevisiae*; the percentage of the proteome predicted to be transmembrane proteins was 25.6% for *Caenorhabditis elegans*, 16% for *Drosophila melanogaster*, and 17.1% for *Saccharomyces cerevisiae*. *Caenorhabditis elegans* and *Drosophila melanogaster* have a similar representation of secreted proteins, while *Saccharomyces cerevisiae* has a significantly lower proportion of these proteins, a finding that may be explained by the fact that *Saccharomyces cerevisiae* is unicellular. Surprisingly, *Caenorhabditis elegans* has nearly double the proportion of transmembrane proteins compared with the other two eukaryotes.

The comparative analysis is carried out using sequence similarity searches against the InterPro database. Data for each of the three complete eukaryotic proteomes are available along with the data for the other complete proteomes. This data, updated weekly, is available at `http://www.ebi.ac.uk/proteome/`. The InterPro database currently enables the characterization of 7293 of the 13,613 *Drosophila melanogaster* proteins (53.6%), 9041 of the 16,606 *Caenorhabditis elegans* proteins (54.4%), and 3231 of the 6174 *Saccharomyces cerevisiae* proteins (52.3%) as belonging to a certain protein family or as possessing a certain domain or functional site. In total, 1673 of the 3208 InterPro entries were found in the three eukaryotic proteomes: 1423 in *Drosophila melanogaster*, 1291 in *Caenorhabditis elegans*, and 1073 in *Saccharomyces cerevisiae*, of which 823 were common to all three species.

Protein kinases belonging to a very extensive family of proteins which share a conserved catalytic core (figure 2.5) with both serine/threonine and tyrosine protein kinases are highly represented in the proteomes all three organisms, accounting for around 2% of the proteome. The C2H2-type zinc finger domain also is abundant in the proteins of all three eukaryotes, making up about 1% of the proteomes of

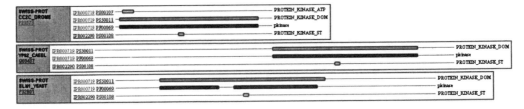

Figure 2.5 InterPro graphical view of representative sequences from the extensive protein kinase family described by InterPro entry IPR000719.

Caenorhabditis elegans and *Saccharomyces cerevisiae*, and accounting for about 2.5% of the proteome of *Drosophila melanogaster*. The high abundance of these protein types in all three eukaryotes would indicate that these proteins are systematically conserved, are likely to have orthologues across species, and are likely to be involved in a shared core biology.

Several of the most abundant families or domains show striking differences in abundance across the three eukaryotic proteomes. The WD repeat is present in a large family of eukaryotic proteins that are implicated in a wide variety of crucial functions (Smith et al., 1999), and the RNA-binding motif, RNP-1, is found in a variety of eukaryotic RNA binding proteins. Proteins of both these types are comparatively underrepresented in *Caenorhabditis elegans*. On the other hand, proteins that belong to the rhodopsin-like G-protein-coupled receptor (GPCR) are unknown in *Saccharomyces cerevisiae*. In fact, only two families are found on all three top 10 lists that number a total of 26 families across the three organisms (table 2.2).

A number of protein types that are apparently unique to a particular species may well define the species. Striking examples are the insect cuticle protein (IPR000618), present only in *Drosophila melanogaster*; the probable olfactory, nematode 7-helix G-protein coupled receptor (IPR000168), present only in *Caenorhabditis elegans* (figure 2.6), and the fungal transcription regulatory protein (IPR001138) and the yeast transposon, Ty (IPR001042), in the *Saccharomyces cerevisiae* proteome. The nematode 7-helix G-protein coupled receptor is the most abundant protein family in the proteome of *Caenorhabditis elegans*, accounting for 3.3% of the proteome.

Together with the rhodopsin-like G-protein-coupled receptor (GPCR) there are a further substantial number of the top ten InterPro families of

Table 2.2 A comparison of the 10 biggest InterPro protein families for *Drosophila melanogaster* versus *Caenorhabditis elegans* and *Saccharomyces cerevisiae*. Only protein kinases (IPR000822) and the C2H2-type zinc finger domain (IPR000276) are found in the top 10 lists of all three organisms

InterPro	D. melanogaster Proteins matched	Rank	C. elegans Proteins matched	Rank	S. cerevisiae Proteins matched	Rank	Name
IPR000822	345	1	188	11	52	9	Zinc finger, C2H2 type
IPR000719	234	2	407	2	118	1	Eukaryotic protein kinase
IPR001254	205	3	13	220	1	926	Serine proteases, trypsin family
IPR001680	175	4	140	15	94	3	G-protein beta WD-40 repeats
IPR002965	175	5	55	51	0		Proline rich extensin
IPR003006	152	6	88	28	1	727	Immunoglobulin and major histocompatibility complex domain
IPR000504	149	7	119	19	56	6	RNA-binding region RNP-1 (RNA recognition motif)
IPR002048	136	8	118	20	19	35	EF-hand family
IPR000379	134	9	116	21	37	14	Esterase/lipase/thioesterase family active site
IPR001611	107	10	54	53	7	132	Leucine-rich repeat

Figure 2.6 InterPro graphical view of the sequence of SRD-1 protein from *Caenorhabditis elegans* (IPR000168).

Caenorhabditis elegans that are present in a much lower percentage in *Drosophila melanogaster* and are absent in *Saccharomyces cerevisiae*. These include the C4-type steroid receptor zinc finger (IPR001628), the C-type lectin domain (IPR001304), the ligand-binding domain of nuclear hormone receptor (IPR000536), and the collagen triple helix repeat (IPR000087) (table 2.3). In contrast, among the top 10 InterPro families

Automation of Protein Sequence Characterization

Table 2.3 The top 10 InterPro families for *Caenorhabditis elegans*. The InterPro entries that have no protein matches against *Saccharomyces cerevisiae* are shown

InterPro	C. elegans		D. melanogaster		S. cerevisiae		Name
	Proteins matched	Rank	Proteins matched	Rank	Proteins matched	Rank	
IPR000168	583	1	0		0		7-Helix G-protein coupled receptor, nematode (probably olfactory) family
IPR000719	407	2	234	2	118	1	Eukaryotic protein kinase
IPR000276	378	3	104	13	0		Rhodopsin-like GPCR superfamily
IPR001810	344	4	23	112	11	59	F-box domain
IPR001628	239	5	22	119	0		C4-type steroid receptor zinc finger
IPR002900	210	6	0		0		Domain of unknown function DUF38
IPR000822	188	7	345	1	52	9	Zinc finger, C2H2 type
IPR001304	188	8	40	52	0		C-type lectin domain
IPR000536	180	9	17	153	0		Ligand-binding domain of nuclear hormone receptor
IPR000561	178	10	95	17	0		EGF-like domain

for *Drosophila melanogaster*, only the proline-rich extensin family (IPR002965) is absent in *Saccharomyces cerevisiae*.

The same domain may be repeated a number of times across a protein sequence. In *Drosophila melanogaster* the type III fibronectin domain (IPR001777) is repeated 39 times in BT gene product (Q9V4F7), the EGF-like domain (IPR000561) is repeated 36 times in N gene product (Q9W4T8), and 28 and 27 times in CG15637 gene product (Q9VR08) and CRB gene product (Q9VC97), respectively. The C2H2-type zinc finger domain (IPR000822) is repeated 23 times in CG11902 gene product (Q9VBR8) and 18 times in CG17390 gene product (Q9V724). In *Caenorhabditis elegans* the immunoglobulin and major histocompatibility complex domain (IPR003006) repeats 48 and 47 times in the UNC-89 gene product (O01761) and hemicentin precursor (O76518), respectively. The low-density lipoprotein (LDL)-receptor class A domain (IPR002172) is the next most common domain with 35 repeats in the low-density lipoprotein receptor-related protein (Q04833), and the EGF-like domain (IPR000561) is repeated 31 times in T20G5.3 gene product (P34576). In *Saccharomyces cerevisiae* the C2H2-type zinc finger domain (IPR000822) is repeated nine times in the transcription factor, TFIIIA (P39933), and seven times in two other proteins (P47043 and P53849).

As mentioned earlier, InterPro gives insights into the domain composition of the classified proteins. *Drosophila melanogaster* has a higher proportion of multidomain proteins (2200) compared with *Caenorhabditis elegans* (1750) and *Saccharomyces cerevisiae* (605). Some *Drosophila melanogaster* multidomain proteins are especially complex, for example, the neutral-cadherin precursor (O15943) and the CG9138 gene product (Q9VM55) each have hits to 11 different InterPro entries. The most complex proteins from *Caenorhabditis elegans* have nine InterPro hits (table 2.4), while the *Saccharomyces cerevisiae* multifunctional carbamoylphosphate synthetase-aspartate transcarbamylase complex (P07259) has seven InterPro hits.

CONCLUSION

Capitalizing on the four-layer strategy for protein analysis described above, it is now possible to perform in-depth whole proteome analysis. For three complete eukaryotic genomes—*Drosophila melanogaster*, *Caenorhabditis elegans*, and *Saccharomyces cerevisiae*—between 52% and 55% of the total proteins are matched in InterPro, and the relative representation of the different protein functions in each set has been assigned using the InterPro functional classification system (figure 2.4). The three proteomes vary in composition, with the emphasis on different cellular functions. For *Saccharomyces cerevisiae*, the largest set of proteins is dedicated to general metabolism, followed by signal transduction, protein synthesis, regulation, and transport. In *Drosophila melanogaster*, metabolism is also the most highly represented function, followed by regulation, signal transduction, and transport. For *Caenorhabditis elegans*, the highest proportion of the proteome is dedicated to regulation, followed by metabolism, signal transduction, GPCRs, and transport. The highly represented protein kinases are represented in the signal transduction category.

The percentage of the proteome dedicated to metabolism increases from *Caenorhabditis elegans* to *Drosophila melanogaster* to *Saccharomyces*

Table 2.4 The 15 proteins from *Caenorhabditis elegans* with the highest number of Inter-Pro hits

Oscode	Protein_ac	Protein_id	Protein name	Hits
CAEEL	Q20204	Q20204	F40E10.4 Protein	10
CAEEL	O61528	O61528	Guanine nucleotide exchange factor UNC-73	9
CAEEL	Q09165	YM01_CAEEL	Hypothetical 1368.6 kDa protein K07E12.1 in chromosome III	9
CAEEL	O44164	O44164	Hypothetical 105.5 kDa protein F16B3.1 in chromosome IV	8
CAEEL	P90891	P90891	F55H12.3 protein	8
CAEEL	Q18990	Q18990	Hypothetical 242.6 kDa protein D2085.1 in chromosome II	8
CAEEL	Q19350	Q19350	Hypothetical 188.4 kDa protein F11C7.4 in chromosome X	8
CAEEL	O02425	O02425	Hypothetical 474.2 kDa protein R31.1 in chromosome V	7
CAEEL	P91904	P91904	Laminin alpha (EPI-1 protein)	7
CAEEL	Q10922	Q10922	Hypothetical 159.2 kDa protein B0286.2 in chromosome II	7
CAEEL	Q20535	Q20535	Similarity to EGF-type repeats	7
CAEEL	Q22070	Q22070	Hypothetical 148.6 kDa protein T01E8.3 in chromosome II	7
CAEEL	Q22098	Q22098	Contains similarity to pfam domain PF00520 (ION_TRANS), score = 887.7, E-value = 1.2E −263, N = 4.	7
CAEEL	Q22275	Q22275	Hypothetical 244.6 kDa protein W07E11.1 in chromosome X	7
CAEEL	Q27512	Q27512	NEX-2 protein	7

cerevisiae, reflecting the greater metabolic diversity of the unicellular organism. *Saccharomyces cerevisiae* also has a higher percentage of proteins dedicated to basic cellular functions, including DNA, RNA, and protein synthesis and the cell cycle. The higher eukaryotes may have evolved more streamlined systems to fulfill these roles. Differences in proteome composition are expected as eukaryotes develop from unicellular organisms to large, multicellular organisms with complex developmental processes.

Rolf Apweiler et al.

REFERENCES

Adams, M. D., Celniker, S. E., Holt, R. A., Evans, C. A., Gocayne, J. D., Amanatides, P. G., Scherer, S. E., Li, P. W., Hoskins, R. A., Galle, R. F., et al. (2000). The genome sequence of *Drosophila melanogaster*. *Science* 287: 2185–2195.

The Arabidopsis Initiative (2000). Analysis of the genome sequence of the flowering plant *Arabidopsis thaliana*. *Nature* 408: 796–815.

Apweiler, R. (2000). Protein sequence databases. In P. Bork (ed.), *Advances in Protein Chemistry*, vol. 54. San Diego and London: Academic Press, pp. 31–71.

Apweiler, R., Attwood, T. K., Bairoch, A., Bateman, A., Birney, E., Biswas, M., Bucher, P., Cerutti, L., Corpet, F., Croning, M. D., Durbin, R., Falquet, L., Fleischmann, W., Gouzy, J., Hermjakob, H., Hulo, N., Jonassen, I., Kahn, D., Kanapin, A., Karavidopoulou, Y., Lopez, R., Marx, B., Mulder, N. J., Oinn, T. M., Pagni, M., Servant, F., Sigrist, C. J., and Zdobnov, E. M. (2001). The InterPro database, an integrated documentation resource for protein families, domains and functional sites. *Nucleic Acids Res*. 29: 37–40.

Ashburner, M., Ball, C. A., Blake, J. A., Botstein, D., Butler, H., Cherry, J. M., Davis, A. P., Dolinski, K., Dwight, S. S., Eppig, J. T., et al. (2000). Gene ontology: Tool for the unification of biology. The Gene Ontology Consortium. *Nat. Genetics* 25: 25–29.

Attwood, T. K., Croning, M. D. R., Flower, D. R., Lewis, A. P., Mabey, J. E., Scordis, P., Selley, J. N., and Wright, W. (2000). PRINTS-S: The database formerly known as PRINTS. *Nucleic Acids Res*. 28: 225–227.

Bairoch, A., and Apweiler, R. (2000). The SWISS-PROT protein sequence database and its supplement TrEMBL in 2000. *Nucleic Acids Res*. 28: 45–48.

Ball, C. A., Dolinski, K., Dwight, S. S., Harris, M. A., Issel-Tarver, L., Kasarskis, A., Scafe, C. R., Sherlock, G., Binkley, G., Jin, H., et al. (2000). Integrating functional genomic information into the Saccharomyces Genome Database. *Nucleic Acids Res*. 28: 77–80.

Bateman, A., Birney, E., Durbin, R., Eddy, S. R., Howe, K. L., and Sonnhammer, E. L. L. (2000). The Pfam Protein Families Database. *Nucleic Acids Res*. 28: 263–266.

Berman, H. M., Westbrook, J., Feng, Z., Gilliland, G., Bhat, T. N., Weissig, H., Shindyalov, I. N., and Bourne, P. E. (2000). The Protein Data Bank. *Nucleic Acids Res*. 28: 235–242.

Blake, J. A., Eppig, J. T., Richardson, J. E., Davisson, M. T., and the Mouse Genome Database Group. (2000). The Mouse Genome Database (MGD): Expanding genetic and genomic resources for the laboratory mouse. *Nucleic Acids Res*. 28: 108–111.

Bork, P., and Koonin, E. V. (1998). Predicting functions from protein sequences—where are the bottlenecks? *Nat. Genetics* 18: 313–318.

The *C. elegans* Sequencing Consortium. (1998). Genome sequence of the nematode *C. elegans*: A platform for investigating biology. *Science* 282: 2012–2018.

Comet, J. P., Aude, J. C., Glemet, E., Risler, J. L., Henaut, A., Slonimski, P. P., and Codani, J. J. (1999). Significance of Z-value statistics of Smith-Waterman scores for protein alignments. *Comput. Chem.* 23: 317–331.

Corpet, F., Servant, F., Gouzy, J., and Kahn, D. (2000). ProDom and ProDom-CG: Tools for protein domain analysis and whole genome comparisons. *Nucleic Acids Res.* 28: 267–269.

Douglas, S. E., and Penny, S. L. (1999). The plastid genome of the cryptophyte alga, *Guillardia theta*: Complete sequence and conserved synteny groups confirm its common ancestry with red algae. *J. Mol. Evol.* 48: 236–244.

Etzold, T., Ulyanov, A., and Argos, P. (1996). SRS: Information retrieval system for molecular biology data banks. *Methods Enzymol.* 266: 114–128.

Fleischmann, R. D., Adams, M. D., White, O., Clayton, R. A., Kirkness, E. F., Kerlavage, A. R., Bult, C. J., Tomb, J. F., Dougherty, B. A., Merrick, J. M., et al. (1995). Whole-genome random sequencing and assembly of *Haemophilus influenzae* Rd. *Science* 269: 496–512.

Fleischmann, W., Moeller, S., Gateau, A., and Apweiler, R. (1999). A novel method for automatic and reliable functional annotation. *Bioinfomatics* 15: 228–233.

The FlyBase Consortium. (1999). The FlyBase Database of the Drosophila Genome Projects and community literature. *Nucleic Acids Res.* 27: 85–88.

Frishman, D., and Mewes, H. W. (1997). PEDANTic genome analysis. *Trends in Genetics* 13(10): 415–416.

Glemet, E., and Codani, J. J. (1997). LASSAP, a LArge Scale Sequence compArison Package. *Comput. Appl. Biosci.* 13: 1317–1343.

Goffeau, A., Aert, R., Agostini-Carbone, M. L., Ahmed, A., Aigle, M., Alberghina, L., Albermann, K., Albers, M., Aldea, M., Alexandraki, D., et al. (1997). The Yeast Genome Directory. *Nature* 387 (suppl.): 5–105.

Hoersch, S., Leroy, C., Brown, N. P., Andrade, M. A., and Sander, C. (2000). The Gene-Quiz Web server: Protein functional analysis through the Web. *Trends Biochem. Sci.* 25: 33–35.

Hofmann, K., Bucher, P., Falquet, L., and Bairoch, A. (1999). The PROSITE database, its status in 1999. *Nucleic Acids Res.* 27: 215–219.

ISO-3309. (1993). Information technology—Telecommunications and information between systems—High-level data link control (HDLC) procedures—Frame structure. In *International Organization for Standardization*, http://www.iso.ch/ 5th ed.

Krause, A., Stoye, J., and Vingron, M. (2000). The SYSTERS protein sequence cluster set. *Nucleic Acids Res.* 28: 270–272.

Nielsen, H., Brunak, S., and von Heijne, G. (1999). Machine learning approaches for the prediction of signal peptides and other protein sorting signals. *Protein Eng.* 12: 3–9.

Oliver, S. G., van der Aart, Q. J., Agostoni-Carbone, M. L., Aigle, M., Alberghina, L., Alexandraki, D., Antoine, G., Anwar, R., Ballesta, J. P., Benit, P., et al. (1992). The complete DNA sequence of yeast chromosome III. *Nature* 357: 38–46.

Rubin, G. M., Yandell, M. D., Wortman, J. R., Gabor Miklos, G. L., Nelson, C. R., Hariharan, I. K., Fortini, M. E., Li, P. W., Apweiler, R., Fleischmann, W., et al. (2000). Comparative genomics of the eukaryotes. *Science* 287: 2204–2215.

Sander, C., and Schneider, R. (1991). Database of homology derived protein structures and the structural meaning of sequence alignment. *Proteins* 9: 56–68.

Sanger, F., Coulson, A. R., Friedmann, T., Air, G. M., Barrell, B. G., Brown, N. L., Fiddes, J. C., Hutchison, C. A., Slocombe, P. M., and Smith, M. (1978). The nucleotide sequence of bacteriophage phi-X174. *J. Mol. Biol.* 125: 225–246.

Scharf, M., Schneider, R., Casari, G., Bork, P., Valencia, A., Ouzounis, C., and Sander, C. (1994). GeneQuiz: A workbench for sequence analysis. In R. Altman, D. Brutlag, P. Karp, R. Lathrop, and D. Searls (eds.), *Proceedings of the Second International Conference on Intelligent Systems for Molecular Biology.* Menlo Park, Calif.: AAAI Press, pp. 348–353.

Smith, T. F., Gaitatzes, C., Saxena, K., and Neer, E. J. (1999). The WD repeat: A common architecture for diverse functions. *Trends Biochem. Sci.* 24: 181–185.

Smith, T. F., and Waterman, M. S. (1981). Identification of common molecular subsequences. *J. Mol. Biol.* 147: 195–197.

Sneath, P. H. A., and Sokal, R. R. (1973). *Numerical Taxonomy*, San Francisco: W. H. Freeman.

Sonnhammer, E. L. L., von Heijne, G., and Krogh, A. (1998). A hidden Markov model for predicting transmembrane helices in protein sequences. In J. Glasgow, T. Littlejohn, F. Major, R. Lathrop, D. Sankoff, and C. Sensen (eds.), *Proceedings of the Sixth International Conference on Intelligent Systems for Molecular Biology.* Vol. 6. Menlo Park, Calif.: AAAI Press, pp. 175–182.

Tamames, J., Ouzounis, C., Casari, G., and Valencia, A. (1998). Automatic classification of proteins in functional classes using database annotations. *CABIOS* 14: 542–543.

Tatusov, R. L., Galperin, M. Y., Natale, D. A., and Koonin, E. V. (2000). The COG database: A tool for genome-scale analysis of protein functions and evolution. *Nucleic Acids Res.* 28: 33–36.

Yona, G., Linial, N., and Linial, M. (2000). ProtoMap: Automatic classification of protein sequences hierarchy of protein families. *Nucleic Acids Res.* 28: 49–55.

SUGGESTED READING

Attwood, T. K. (2000). The role of pattern databases in sequence analysis. *Briefings in Bioinformatics* 1: 45–59. A well-written review focusing on databases of protein signatures for families and domains. Provides an excellent overview of the current status of these databases, outlines the methods behind them, and discusses their diagnostic strengths and weaknesses.

Apweiler, R. (2001). Functional information in SWISS-PROT: The basis for large-scale characterisation of protein sequences. *Briefings in Bioinformatics* 2: 9–18. Describes the essential role that the SWISS-PROT protein sequence database plays in automatic large-scale functional characterization and annotation due to its high level of functional information.

Apweiler, R., Attwood, T. K., Bairoch, A., Bateman, A., Birney, E., Biswas, M., Bucher, P., Cerutti, L., Corpet, F., Croning, M. D. R., Durbin, R., Falquet, L., Fleischmann, W., Gouzy, J., Hermjakob, H., Hulo, N., Jonassen, I., Kahn, D., Kanapin, A., Karavidopoulou, Y., Lopez, R., Marx, B., Mulder, N. J., Oinn, T. M., Pagni, M., Servant, F., Sigrist, C. J. A., and Zdobnov, E. M. (2001). *Nucleïc Acids Res.* 29: 37–40. This paper describes InterPro, the integrated documentation resource of protein families, domains, and functional sites that currently combines the complementary efforts of the PROSITE, PRINTS, Pfam, ProDom, and SMART database projects. InterPro contains manually curated documentation, combined with diagnostic signatures from different databases allowing users to see at a glance whether a particular family or domain has associated patterns, profiles, fingerprints, etc. In the task of sequence characterization, InterPro will provide a more reliable, concerted method for identifying protein family traits and for inheriting functional annotation. This is especially important given the dependence on automatic methods for assigning functions to the raw sequence data issuing from genome projects.

Apweiler, R., Biswas, M., Fleischmann, W., Kanapin, A., Karavidopoulou, Y., Kersey, P., Kriventseva, E. V., Mittard, V., Mulder, N., Phan, I., and Zdobnov, E. (2001). Proteome Analysis Database: Online application of InterPro and CluSTr for the functional classification of proteins in whole genomes. *Nucleic Acids Res.* 29: 44–48. A description of the Proteome Analysis Database (`http://www.ebi.ac.uk/proteome/`), which provides comprehensive statistical and comparative analyses of the predicted proteomes of fully sequenced organisms. The analysis is compiled using InterPro (`http://www.ebi.ac.uk/interpro/`), CluSTr (`http://www.ebi.ac.uk/clustr/`), and GO (`http://www.geneontology.org/`), and is performed on nonredundant sets of SWISS-PROT and TrEMBL entries representing each complete proteome. The Proteome Analysis Database currently contains statistical and analytical data for the proteins from 45 complete genomes, including five eukaryotes: *Arabidopsis thaliana, Caenorhabditis elegans, Drosophila melanogaster, Homo sapiens*, and *Saccharomyces cerevisiae*.

Ashburner, M., Ball, C. A., Blake, J. A., Botstein, D., Butler, H., Cherry, J. M., Davis, A. P., Dolinski, K., Dwight, S. S., Eppig, J. T., Harris, M. A., Hill, D. P., Issel-Tarver, L., Kasarskis, A., Lewis, S., Matese, J. C., Richardson, J. E., Ringwald, M., Rubin, G. M., and Sherlock, G. (2000). Gene ontology: Tool for the unification of biology. The Gene Ontology Consortium. *Nat. Genetics* 25: 25–29. The authors describe the creation of a dynamic, controlled vocabulary that can be applied to all eukaryotes even as knowledge of gene and protein roles in cells is accumulating and changing. To this end, the Gene Ontology (GO) Consortium has constructed three independent ontologies: biological process, molecular function, and cellular component. The GO concept is intended to make possible, in a flexible and dynamic way, the annotation of homologous gene and protein sequences in multiple organisms, using a common vocabulary that results in the ability to query and retrieve genes and proteins based on their shared biology. The GO ontologies produce a controlled vocabulary that can be used for dynamic maintenance and interoperability between genome databases.

URLs FOR RELEVANT SITES

EBI home page. `http://www.ebi.ac.uk/`

SWISS-PROT and TrEMBL protein sequence databases.
`http://www.ebi.ac.uk/swissprot`

Sequence Retrieval System (SRS). `http://srs.ebi.ac.uk/`

InterPro. `http://www.ebi.ac.uk/interpro/`

CluSTr. `http://www.ebi.ac.uk/clustr/`

Proteome Analysis Database. `http://www.ebi.ac.uk/proteome/`

Proteome analysis data for human. `http://www.ebi.ac.uk/proteome/HUMAN/`

Program to run interactive proteome analysis for user-selected organisms.
`http://www.ebi.ac.uk/proteome/comparisons.html`

Information on and how do download the nonredundant complete proteome sets maintained at the EBI. `http://www.ebi.ac.uk/proteome/CPhelp.html`

PDB-Protein Data Bank. `http://pdb-browsers.ebi.ac.uk/`

OTHER USEFUL URLs

GO (the Gene Ontology Consortium). `http://www.geneontology.org/`

Methods for protein annotation.
`http://jura.ebi.ac.uk:8765/ext-genequiz/`
`http://pedant.gsf.de/`

Clustering methods. `http://www.ncbi.nlm.nih.gov/COG/`
http://www.protomap.cs.huji.ac.il/
http://www.dkfz-heidelberg.de/tbi/services/cluster/systersform

HSSP database, a database of homology-derived secondary structure of proteins for classification purposes. `http://www.sander.ebi.ac.uk/hssp/`

SignalP, which predicts the presence and location of signal peptide cleavage sites in amino acid sequences. `http://www.cbs.dtu.dk/services/SignalP/`

TMHMM, for prediction of transmembrane helices in proteins.
`http://www.cbs.dtu.dk/services/TMHMM-2.0/`

3   Information Fusion and Metabolic Network Control

Andreas Freier, Ralf Hofestädt, Matthias Lange, and Uwe Scholz

Molecular biology, biotechnology, and bioinformatics have started to focus on the problem of gene-regulated metabolic network control. This problem cannot be circumvented, because no open reading frame can be expressed without the appropriate regulatory sequences. Moreover, some genes code for proteins that turn other genes on and off. Groups of these genes constitute networks with complex behaviors. These networks control other genes whose protein products catalyze specific biochemical reactions. Hence the small molecules that are substrates or products of these reactions can in turn activate or deactivate proteins that control transcription or translation. Therefore, gene regulation can be said to indirectly control biochemical reactions in cellular metabolism, and cellular metabolism to exert control of gene expression.

For these reasons, the interdependent biochemical processes of metabolism and gene expression can, and should, be interpreted and analyzed in terms of complex dynamical networks. Therefore modeling and simulation are necessary. To solve this problem, we have to bring together information from gene regulation and metabolic pathways. These data have been, and will be, stored systematically in specific databases that are accessible via the Internet. Recently many firms have been established that provide essential data for the solution of scientific and industrial problems and, even more important, the corresponding infrastructure. As a result, there are more than 200 databases available via the Internet for all known sequenced genes (e.g., EMBL), proteins (e.g., SWISS-PROT, PIR, BRENDA), transcription factors (e.g., TRANS-FAC), biochemical reactions (KEGG), and signal induction reactions (e.g., TRANSPATH, GeneNet). Besides databases, simulators for metabolic networks, which employ most of the currently popular modeling methods, are also available via the Internet. In addition to the classical methods of differential equations, discrete methods have become quite important. The integration of relevant molecular database systems and the analysis tools will be the backbone of powerful information systems for biotechnology.

This chapter introduces the basics of database and information systems for molecular biology. Based on this introduction, we will discuss the important topics of database integration for the automatic fusion of molecular knowledge. Furthermore, computer science is developing and implementing tools for the analysis of molecular data. The analysis of molecular data and of metabolic networks are of equal importance. The chapter presents the rule-based model of metabolic processes, which is the core of our simulation shell, and also shows the prototype of the MARG-Bench system (MARG, Modeling and Animation of Regulative Gene Networks), which integrates basic molecular database systems and the simulation tool.

DATABASES AND INFORMATION SYSTEMS

For the physical storage and maintenance of data by means of computers, specific software is necessary. In the simplest case, the functions of the operating system are used and the data are stored in simple files. Therefore the biological software tools themselves specify the internal structure and sequentially write the data into related files. The tools are as different as the file formats used. These range from unreadable binary format to well standardized XML (Extensible Markup Language). In this respect the software engineer of each tool is responsible for the data storage and must implement methods for data access, updates, and backup (see figure 3.1).

This kind of data storage and access is often mistaken for a database management system (DBMS). Some important disadvantages of such cumulatively acquired data collections are lack of performance, data redundancies, and lack of standardized query languages, synchronous data updates, and explicit scheme information that can be called up. These disadvantages are often unimportant for individual uses, but in the case of multiple tool usage and massive global and public use, they present a problem. Nevertheless, the necessity of their use is generally accepted, because all collected data have to be analyzed, using several

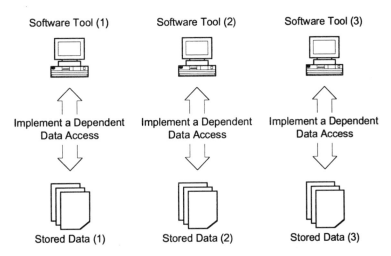

Figure 3.1 Schematic of individual dependent data access.

tools and techniques. It is important to enable efficient data access that is independent from the implementation of the storage mechanism (see figure 3.2).

Consequently, for a common and wide-ranging use of data, a homogeneous kind of data storage and access is essential. Therefore the main tasks of a DBMS are the following (Codd, 1982):

- Uniform management of all data
- Provision of data operations such as storage, update, and search
- Covering a unique data description of all stored data (scheme)
- Consistency check
- Authorization control
- Transaction management
- Backup mechanism.

Today many DBMS are available that differ in price, data model (hierarchical, relational, object-oriented, object-relational etc.), features (supported operators, transactions, indexes etc.), and query mechanism and languages (SQL, OQL, QBE, etc.). Popular examples of such DBMS are Oracle, INFORMIX, DB2, Microsoft SQL Server, POET, and Object Store.

The main advantages of such DBMS are very efficient searching and access opportunities, common usable data, few data storage redun-

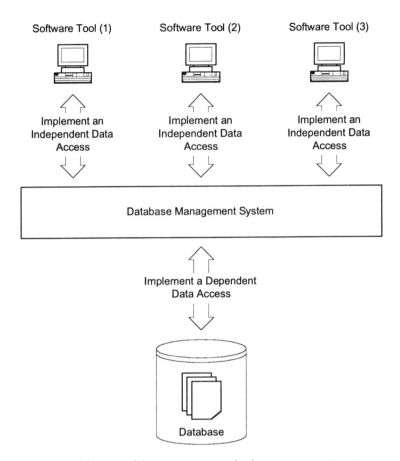

Figure 3.2 Schematic of data access using a database management system.

dancies, and easy data access by software tools. Thus, DBMS are increasingly used for biological data (Kemp et al., 1999; Xie et al., 2000).

In most cases stored biological data are not independent. Rather, they are processed by related biological tools. That means that tools and related data represent a coupled system, which is called an information system (IS). An IS is a complex, coupled software system for information processing. In other words, we can define an IS as a logical union of the data itself, tools, and the import of external data. Consequently, a DBMS may be identified as the central substructure of persistent data storage of such a system.

Andreas Freier et al.

MOLECULAR DATABASES

In molecular biology a large number of research projects are currently producing an exponentially increasing amount of data. A popular example is the Human Genome Project. In this context alone about 3×10^9 base pairs and mapping data must be stored (see `http://www.ornl.gov/hgmis/project/progress.html`). The most popular public sequence database, EMBL, includes 4.7 million entries of primary nucleotide sequences and related data (Baker et al., 2000). If these databases are set in relation to the count of molecular databases, an impression of the data volume may be given. These data are often published in publicly accessible sources. Often the categories are very different. One investigation has divided sequence and related data into 16 categories (Burks, 1999). In correspondence with this, about 200 WWW-based data sources are listed. This is meant to be only a fraction of the overall number of databases available. In most cases these data are stored for processing, analyzing, and other functions. A common approach is to collect all available data, then decide which are dispensable. However, storage capacity is not a problem today.

Nowadays the majority of the molecular database providers store their research results in "quick-and-imperfect" systems. In most cases simple flat files are used, which are managed by the directory and file system structures of the operating system. Data access is achieved with software that can handle the individual file format. As an interface for external access, use of the WWW is popular. In those cases HTML forms coupled with CGI scripts are used, and the data obtained by this method are presented in HTML pages, using tables or other formats. A popular example is the KEGG system (Ogata et al., 1999).

One of the reasons why the content of these databases constantly increases is clarified in a special issue of *Nucleic Acids Research* that is published annually in January and gives an overview of databases as well as presenting selected systems in more detail. In order to provide a rough picture of the extent of currently available databases, a selection of important databases is listed in the appendix. A quite detailed collection on this topic has been compiled under the following URL: `http://www-bm.cs.uni-magdeburg.de/iti_bm/marg/dataacquisition/data_sources.html`. The most important systems in the field of molecular network control are listed below.

Genes

- EMBL (`http://www.ebi.ac.uk`)
- GDB (`http://gdbwww.gdb.org/`)
- GenBank (`http://www.ncbi.nlm.nih.gov`)

Proteins and Enzymes

- ENZYME (`http://www.expasy.ch`)
- LIGAND (`http://www.genome.ad.jp`)
- PDB (`http://www.pdb.bnl.gov/`)
- PIR (`http://pir.georgetown.edu`)
- SWISS-PROT (`http://expasy.hcuge.ch/sprot/sprot-top.html`)

Pathways

- ExPASy (`http://www.expasy.ch`)
- KEGG (`http://www.genome.ad.jp`)
- WIT (`http://wit.mcs.anl.gov/WIT2`)

Gene Regulation

- EPD (`ftp://ftp.ebi.ac.uk/pub/databases/epd/`)
- RegulonDB (`http://www.cifn.unam.mx/Computational_Genomics/regulondb/`)
- TRANSFAC (`http://transfac.gbf.de/`)
- TRRD (`http://www.bionet.nsc.ru/SRCG/index.html`)

DATABASE INTEGRATION AND INFORMATION FUSION

In order to use these special databases, the user has to connect to each database separately. However, the integration of databases can help in detection of new information. An example is the interaction between aligned gene sequences of an organism to supplement an unknown part of a metabolism or suggest and predict alternative pathways

(Luttgen et al., 2000). At first sight, two basic problems may arise when handling such a distributed data retrieval:

• How does the user access each relevant database?
• How can the several query results be merged into a joined data set?

These problems lead directly to the field of database integration. The integration of databases has several parts.

Data merging is the overcoming of the distributed data storage. On the one hand, the data overlap between the several databases, and on the other hand, the data are totally different. To produce a global data set, it becomes necessary to implement a unique global data store. There are two different ways to achieve this: copying all needed data into one large database (materialized) and leaving the data where they are and merging them virtually (nonmaterialized).

Derivation of an integrated scheme starts from the fact that every local database provides its own data scheme. But in the case of global access, a unique scheme over all databases is needed. The method used to solve this problem specifies the degree of scheme integration. Consequently, it is possible to define one global scheme over all local schemes, meaning that all locally modeled data are integrated into a global scheme (bottom up). The other kind of scheme integration is to specify which data are needed by individual application scenarios, thus making it possible to model several partly integrated global schemes (top down).

For adequate access to the integrated data, a unique data access and querying method is needed. For this, the mapping of local data models and access methods to global ones has to be carried out. In the case of read-only data access, this is no problem, but for writing operations, problems occur that include transactions of global management methods for simultaneous writing on the same object, data consistency mechanisms, and global integrity policies.

Besides the data retrieval, the quality of the integrated data is important. Therefore, it must be determined how reliable the data actually are (Bork and Bairoch, 1996). It is generally accepted that databases include faulty or low-quality entries such as incorrect sequences, missed annotation, and wrongly assigned enzyme numbers. Consequently, these quality problems are propagated during the database integration to the global data set. Hence, a mechanism for quality control of the databases and their entries must be established.

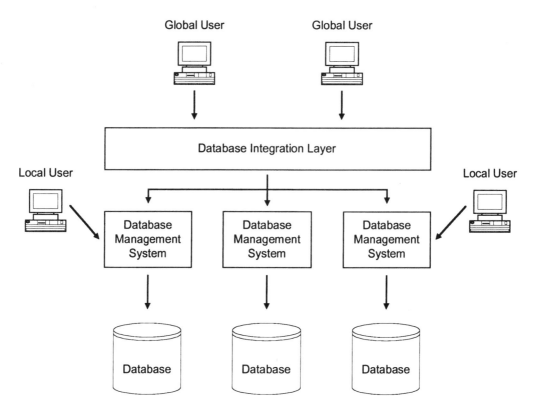

Figure 3.3 Integrative data access using database integration software.

A software layer has to be provided that offers methods for data integration. Figure 3.3 illustrates this situation.

Approaches for software layers deal with such problems as inadequate support of standardized query languages, nonappropriate programming interfaces, and insufficient consideration of individual user requirements. These problems catalyzed the implementation of an alternative approach, called MARG-Bench, which is described later in this chapter.

Moreover, new concepts of "information fusion" have been developed that are focused on "information discovery" and on the basis of database integration. The notion of transformation includes integration, filtering, analysis, and preparation of data aimed to discover and represent the hidden knowledge.

Related Work

The main difficulties are the unique access to proprietary flat files, Internet interfaces, and the large amount of data. Several ways of realizing database integration for biological data exist and are currently in use:

• Hypertext Navigation: KEGG (Ogata et al., 1999)

• Data Warehouse: SRS (Etzold et al., 1996), PEDANT (Frishman and Mewes, 1997), HUSAR (Senger et al., 1995)

• Multi Database Queries: BioKleisli (Davidson et al., 1997), OPM (Topaloglou et al., 1999).

• Mediator techniques: Multiagents (Matsuda et al., 1999).

One of the most developed technologies of web integration of molecular databases uses SRS (Sequence Retrieval System (Etzold et al., 1996). It is based on local copies of each component database, which have to be provided in a text-based format. The results of the query are sets of Internet links through which the user can navigate. As of 2000, more than 100 databases on molecular biology are integrated under SRS. However, within the limitations of this approach, data fusion is still the task of the user. Real data fusion (i.e., data for one real-life object, such as an enzyme) from two different databases (e.g., KEGG and BRENDA) is not found to be represented twice by different web pages.

Therefore, research groups try to integrate molecular databases on a higher level than the SRS approach. Results of current database research, including federated database systems, data warehousing architectures, and data mining techniques (Conrad, 1997), are applied. Many problems in bioinformatics require the following:

1. Access to data sources that are large in volume, highly heterogeneous and complex, constantly evolving, and geographically dispersed

2. Solutions that involve multiple, carefully sequenced steps

3. Smooth passing of information between the steps

4. An increasing amount of computation

5. An increasing amount of visualization.

BioKleisli (Davidson et al., 1997) is an advanced technology designed to handle the first three requirements directly. In particular, it provides the high-level query language, CPL, that can be used to express complicated transformation across multiple data sources clearly and simply. In addition, while BioKleisli does not handle the last two requirements directly, it is capable of distributing computation to appropriate servers and initiating visualization programs.

The Idea of the Integration Approach

In consideration of the advantages of integrative data access and the existence of inflexible approaches, a new system for database integration was developed (see `http://www-bm.cs.uni-magdeburg.de/iti_bm/marg/`). According to the latest findings, the system offers a scalable and flexible approach. This is achieved by the concept of wide-ranging database access by a wrapper (adapter) technology. The data merging is nonmaterialized, using set operations. Specific data integration-related schemes for the user can be defined, and the standardized access to the integrated data can be performed by an SQL-like language. For comfortable use, a JDBC driver is available.

MODELING OF METABOLIC NETWORKS

Using the rule-based modeling of metabolic processes, the simulation environment MetabSim for the analysis and visualization of gene-controlled metabolic processes was implemented. The advantage of this concept is the integration of relevant molecular database systems that are available via the Internet.

Related Work

The availability of the rapidly increasing volume of molecular data on genes, proteins, and metabolic pathways improves the capability to study cell behavior. To understand the molecular logic of cells, it must be possible to analyze metabolic processes and gene networks in qualitative and quantitative terms. Therefore, modeling and simulation are important methods.

Mathematical models may be subdivided into two categories: analytical and discrete. Analytical models perform the processes of elements, acting as functional relations (algebraic, integral-differential, finite-differential, etc.) or logical conditions. An analytical model may

Andreas Freier et al.

be studied by qualitative, analytical, or numerical methods. Analytical models are generally based on integral and differential systems of equations. The paper published by Waser et al. (1983) presents a computer simulation of phosphofructokinase. This enzyme is a part of the glycolysis pathway. Waser and coworkers model all kinetic features of the metabolic reaction, using computer simulation. This computer program is based on the rules of chemical reaction, which are described by differential equations. Franco and Canelas (1984) simulate purine metabolism by differential equations; each reaction is described by the relevant substances and the catalytic enzymes, using the Michaelis constant of each enzyme.

Discrete models are based on state transition diagrams. Simple models of this class are based on simple production units, which can be combined. Overbeek (1992) presented an amino acid production system. A black box with an input set and an output set displays a specific production. The graphical model of Kohn and Letzkus (1982), which allows the discussion of metabolic regulation processes, is representative for the class of graph theoretical approaches. Kohn and Letzkus expand the graph theory by a specific function that allows the modeling of dynamic processes. In this case, the approach of Petri nets is a new method. Reddy et al. (1993) presented the first application of Petri nets in molecular biology. This formalism is able to model metabolic pathways (Hofestädt and Thelen, 1998). The highest abstraction level of this model class is represented by expert systems and object-oriented systems (Brutlag et al., 1991; Stoffers et al., 1992). Expert systems and object-oriented systems are developed by higher programming languages (Lisp, C++) and allow the modeling of metabolic processes by facts/classes (protein and enzymes) and rules/classes (chemical reactions).

The grammatical formalization is able to model complex metabolic networks. Within this class of models one may consider the cell model (E-CELL), developed by Tomita et al. E-CELL is the generic computer software environment for modeling and simulation of whole cell systems (see Tomita et al., 1999; see also chapter 11). It is an object-oriented environment for the simulation of molecular processes in user-definable models, equipped with interfaces that allow observation and intervention, and written in C++. Using E-CELL, one could construct a hypothetical cell with a definite number of genes sufficient for transcription, translation, energy production, and phospholipid synthesis.

Rule-Based Modeling

The model is an extension of the Semi-Thue system. Using a universal rule, this formalization allows the representation of genetic, biosynthetic, and cell communication processes. Furthermore, it is necessary to expand this discrete model by adding concentration rates for each metabolite. Metabolites are substances or substance concentrations that can be modified by biochemical reactions. Enzymes are specific proteins that catalyze biochemical reactions. Inducers and repressors are metabolites that can accelerate or slow down/prevent biochemical reactions. The biochemical space (cell state) of a cell is a mixture of these components. The set of all cell states will be denoted by Z. By these definitions the abstract metabolism is defined by the actual cell state and the biochemical reaction rules. The metabolic rule is the basic unit of the metabolic system.

Let Z be a finite set of cell states. A 5-tuple (B, A, E, I, p) with $p \in [0, 1]_{\mathbb{Q}}$ and $B, A, E, I \in Z$ is called a metabolic rule. p is rule probability; B (Before) is a set of preconditions; A (After) is a set of postconditions; E (Enzyme) is a set of catalyzed conditions; and I (Inhibitor) a set of inhibitor conditions.

Example 1 The reversible biosynthesis product glucose-6-phosphate $\leftrightarrow$ fructose-6-phosphate will be catalyzed by the enzyme glucosephospat-isomerase. This process can be described by two rules:

- ({glucose-6-phosphate}, {fructose-6-phosphate}, {glucosephosphate-isomerase}, { }, p)
- ({fructose-6-phosphate}, {glucose-6-phosphate}, {glucosephosphate-isomerase}, { }, p).

On the basis of the metabolic rule, the basic model can be defined.

$G = (Z, R)$ is a metabolic system. Z is a finite set of cell states, $S \in Z$ is the start state, and R is the metabolic rule set.

In the following paragraph, the semantics of the metabolic system are defined. The integration of the analyzed metabolic features is the basis of this formalization and is the reason for specifying a stochastic parallel derivation mechanism that will describe the change of actual cell states, depending on the specified rule set. Therefore, the set of all activated rules must be fixed. This is the first step of the derivation process. A rule is "activated" if its preconditions are elements of the actual state

Andreas Freier et al.

$z \in Z$. Moreover, effects of inducer and inhibitor elements must be considered. If such metabolites are elements of the actual state z, then the probability of this rule will be modified by inhibitor and inducer effects. The special function CALCULATE(z, r) will determine the absolute probability value as rule r depending on state z. A random generator (RANDOM), using the absolute probability value of the input, works as a Boolean function and will produce either positive or negative results (true or false). Once the Boolean value is true (false), rule $r \in R$ is described as activated (deactivated) and goes into action.

Let $G = (Z, R)$ be a metabolic system, $r = (B, A, E, I, p) \in R$ a rule, and $z \in Z$ a cellular state. r will be activated by z (r_z), iff $\forall x \in B, x \in z$. $A(z) = \{r \in R : r \text{ is activated by } z\}$ is the set of rules activated by z.

Example 2 Let $G = (Z, R, z_0)$ be a metabolic system and $z_0 = \{S, D\}$ and $R = \{r_1, r_2\}$ with

$r_1 = (\{S\}, \{H, S\}, \{D\}, \{L\}, 0.8)$

$r_2 = (\{D\}, \{X\}, \{E\}, \{D\}, 0.6)$

For the configuration z_0 $A(z_0) = \{r_1\}$ the rule r_2 is not activated because the repressor is available.

Any activated rule $r \in R$ can go into action. The action of r will modify the actual cellular state of the metabolic system. Elements of the actual cellular state, which are elements of the Before set of rule r, will be eliminated in z and all elements of the After component will be added to z. Therefore, the action of rule r can produce a new state, $z' \in Z$.

Let $G = (Z, R)$ be a metabolic system, $z \in Z$ the actual cellular state, and $r_z = (B, A, E, I, p) \in R$. The action of r_z is defined thus: If RANDOM(CALCULATE(z, r)) = true, then $z' = (z - B) \cup A$. The action of r_z will be written as $z \rightarrow_r z'$.

According to the metabolic system defined in example 2, the action of r_1 will produce the state $z' = \{H, S, D\}$.

The one-step derivation of a metabolic system is defined by the (quasi) simultaneous action of all activated rules. Therefore, the set of all activated rules is considered and two new sets are determined: the Before set and the After set. The Before set includes all B elements of the activated rules. The definition of the After set is analogous. Using these sets, the one-step derivation can be interpreted as an addition and subtraction of elements.

Example 3   Let $G = (Z, R, z_0)$ be a metabolic system and $z_0 = \{B, C, E\}$ and $R = \{r_1, r_2, r_3\}$, with

$r_1 = (\{B\}, \{S, B\}, \{C\}, \{\ \}, 0.9)$

$r_2 = (\{C\}, \{F, C\}, \{E\}, \{\ \}, 0.3)$

$r_3 = (\{C, F\}, \{B, C\}, \{E\}, \{\ \}, 0.3)$.

$A(z_0) = \{r_1, r_2\}$ is the set of activated rules. For z_0 we can identify one-step derivations that will produce the following states:

$\{B, C, E, S\}$ (action of r_1)

$\{B, C, E, F\}$ (action of r_2)

$\{B, C, E, S, F\}$ (action of r_1 and r_2)

$\{B, C, E\}$ (empty action).

Let $G = (Z, R)$ be a metabolic system, $z \in Z$ the actual cellular state, $A(z)$ the set of all activated rules under z, and $B_z = \{B : \exists r \in A(z) \wedge B \in r_z\}$ and $A_z = \{A : \exists r \in A(z) \wedge A \in r_z\}$. The simultaneous action of $A(z)$ is called one-step derivation if $z' = (z - B_z) \cup A_z$. It is written $z \Rightarrow z'$.

Each action can be interpreted as an independent event. Therefore, the probability of each one-step derivation can be calculated in terms of the absolute probability values of all activated and deactivated rules. In the simulation system, this will be done by multiplying these values.

However, based on the one-step derivation, the derivation can be defined inductively, and a probability for any derivation can be calculated.

Let $G = (Z, R)$ be a metabolic system. $x \in Z^+$ is a derivation in G iff $|x| = 1$ or $|x| > 1$; $\exists y' \in Z^* z'$, $z'' \in Z : x = z' z'' y$ and $z'' y$ is a derivation; and $z' > z''$.

Example 4   Let be G the metabolic system defined in example 3. The following sequence of cellular states describes a derivation:

$\{B, C, E\} \Rightarrow T \ \{B, C, E, S\} \Rightarrow T' \ \{B, C, E, S, F\} \Rightarrow T'' \Rightarrow \{B, C, E, S, F\} \ \cdots$ where $T = \{r_1\}$, $T' = \{r_1, r_2\}$, and $T'' = \{r_2, r_3\}$.

In the case of analytical modeling, it is necessary to expand the model, using abstract concentration rates. To realize this requirement for each component (metabolite), specific integer values must be assigned. These values can be interpreted as abstract concentration rates. Since these impacts can be included in the metabolic system, us-

Andreas Freier et al.

ing the formalization of multi-sets, the definition of the metabolic system must be modified. Regarding the rule activation, the concentration rate of any Before component must be satisfied in regard to corresponding metabolites of the actual state. The concentration rate of these metabolites must be higher than or equal to the corresponding Before component of this metabolic rule.

Moreover, the function CALCULATE must be modified. In this case, the influence of all concentration rates of inductor and repressor metabolites will determine the absolute rule probability. All activated rules can go into action simultaneously. With regard to corresponding metabolites of the actual state (integer values), the addition and subtraction of the concentration rates of all Before and After components is necessary. In this chapter, only the fundamental part of this formalization is presented.

Let $z \in Z$ be a state. The multi-set $k : z \to \mathbb{Z}_0$ is metabolic concentration. K denotes the set of all metabolic concentrations.

Example 5 Let $z = \{$glucose, lactose, RNA-polymerase$\}$. [34 glucose, lactose, 15 RNA-polymerase] defines a metabolic concentration of 34 molecules glucose, one molecule lactose, and 15 molecules RNA-polymerase.

Based on the formalization of multi-sets, the analytical metabolic system, which enables the discussion of kinetic effects, can be defined. $G = (Z, R, k)$ with $A \in Z$ the start state, $k \in K$ a multi-set $(K : A \to \mathbb{N})$, and R a finite set of metabolic rules, where $r = (B, A, E, I, p) \in R$ with $p \in [0, 1]_{\mathbb{Q}}$ and B, A, E, I are metabolic concentrations, is the metabolic system.

The definition of activation is necessary. Regarding multi-sets, the activation of any rule $r \in R$ depends on the specified concentration rates of each rule component, the concentration rates of the actual state, and the absolute rule probability.

Let $G = (Z, R, k)$ be an analytical metabolic system, $r \in R$ a rule, and $z \in Z$ a cellular state. r is activated by z (r_z), iff $\forall x \in B$, $x \in z$, and $k(x) \le k(z_x)$. $A(z) = \{r \in R : r$ is activated by $z\}$ is the set of rules activated by z.

Based on this definition, the one-step derivation can be modified. All activated metabolic rules can go into action simultaneously. Regarding the set of activated rules, all After concentration rates must be added to the actual state and all Before concentration rates must be subtracted from the actual state.

$G = (Z, R, k)$ is an analytical metabolic system; z, the actual cellular state; $A(z)$, the set of all activated rules under z; and

$$B_z = \{B : \exists r \in A(z)\ B \in r_z\} \quad \text{and} \quad |B_z| := \sum_{b \in B} k(b)$$

$$A_z = \{A : \exists r \in A(z)\ A \in r_z\} \quad \text{and} \quad |A_z| := \sum_{a \in A} k(a).$$

The simultaneous action of $A(z)$ is called one-step derivation iff $z' = \{x : x \in A_z \text{ or } \forall x \in z, y \in B_z x = y \text{ and } k(x) - k(y) > 0\}$. The one-step derivation is written $z \to z'$.

Using the one-step derivation operator, one can define the derivation of an analytical metabolic system inductively.

Example 6 Let G be an analytical metabolic system with $k = [6A, 8B, 3E]$ and the rule set R with

$r_1 = ([2A], [3B], [E], [X], 0.8)$

$r_2 = ([2B], [3D], [E], [X], 0.5).$

$A(k) = \{r_1, r_2\}$ and the set $\{[6A, 8B, 3E], [4A, 11B, 3E], [6A, 6B, 3D, 3E], [4A, 9B, 3D, 3E]\}$ describes the different one-step derivations.

Application

The implemented universal metabolic rule allows the formalization of biosynthesis, gene expression, gene regulation, and cell communication processes. Regarding biosynthetic processes, the before, after, inducer, and repressor components are used. For example, Enzyme E_1 will catalyze the biochemical process S_1 into S_2. This can be expressed by $B = S_1$, $A = S_2$, and $E = E_1$. Moreover, one can add any concentration rates. For example, $B = 15S_1$, $A = 12S_2$, $E = 2E_1$. The probability value will model the flux of this biochemical rule, which can be influenced by specific inducer and repressor metabolites, depending on their concentration rates.

In the case of simple cell communication processes, only the before or after component will be used. From this, we obtain the following interpretation: Substance A enters the cell by endocytotic processes. Therefore, a rule must be defined in which only the after component will be assigned specific substances. Moreover, such processes can be

Andreas Freier et al.

influenced by specific events that can be formulated regarding inducer and repressor components.

Normally, metabolites will disintegrate after a specific time interval. This can be expressed by a rule that represents only the specific before component. Moreover, concentration rates and specific influence components can be defined.

In the case of gene regulation, the activity of operons can easily be modeled. If we choose an operon that represents two structure genes (S_1, S_2), two operator genes (O_1, O_2), and one enhancer sequence, then this can be expressed by

$B =$ [RNA-polymerase, ribosome, amino acid, tRNA]

$A = [S_1, S_2], \quad E = [IO_1, IO_2] \quad$ and $\quad I = [O_1, O_2].$

However, this model allows the simulation of complex metabolic networks, and the grammatical formalization allows the definition and implementation of different languages. These languages represent specific biological aspects. For example, it is possible to produce the set of all possible pathways, to produce metabolic pathways depending on specific conditions (for example, the probability value), to search for the appearance of specific substances (for example, toxic substances), and so on.

Theoretical Aspects

The set of the attainable configurations is an infinite set, and the set of all derivations is enumerable. Moreover, the set of all configurations is undecidable (Hofestädt, 1996). Use of concentrations (multi-sets) is the main reason for the undecidability. However, this result implies that no interesting question can be solved in the field of biotechnology research.

In practice, biochemical systems are restricted. In the model, the depth and width of the derivation process can be restricted. Therefore, important questions are decidable, and the complexity of the derivation algorithm must be discussed. If the derivation depth is restricted, the language $L(G, i)$ is decidable. $L(G, i)$ is the set of all configurations that can be produced from the start configuration by the application of up to i derivation steps. Hence, for a metabolic system with the generation depth i, it is decidable if k is a member of $L(G, i)$. Based on the exponential complexity of the derivation process, this question cannot be solved in practice if i is high. Thus the calculation of a derivation tree is not possible in practice because the one-step derivation represents an exponential time complexity. However, the simulation tool is used, the derivation depth and/or width must be restricted.

MetabSim

The simulation system MetabSim is the current implementation of the rule-based model described above. It consists of two main parts. The first part, the object-oriented data structure, describes the metabolic grammar. The main structure here is the metabolic rule. A rule contains the stoichiometry of substrates and products, enhancers, inhibitors, and factors, and the elasticity coefficients of one complex reaction. The user can instantiate the rule type to build up rules according to reactions of a metabolic network. The second data type is the metabolic state. The user can define a metabolic state to act as a root configuration. All later calculations will be based on this state. The whole data structure is mapped into a database so that all rules and states are stored therein.

The second part of MetabSim is the derivation logic. Because the system has been designed to be modular, several derivation modules can be implemented and applied independently. Figure 3.4 shows the information flow in the rule-based simulation system. After the rule set and the root configuration (default cell states) are defined, the derivation logic can be applied to the data. In the first step, the Rule Selection module accesses the current state and calculates the rules that can be applied because their premises are true in relation to the current state. The Rule Application module calculates the successive configuration(s). Optionally, the reaction time is applied by a Rule Kinetics module. The new configurations (states) are the input for the next derivation step.

In biochemistry, for a set of single metabolic systems, the systems behavior is described by the use of differential equations. But the problem is that this approach does not discuss all interactions in large networks. It was decided to apply a simpler and more abstract method as a compromise between the mathematical modeling and the data outcome of molecular databases. The BRENDA database stores kinetic parameters of enzymes. From this system one can get the Michaelis-Menten value, the turnover rate, and the type of regulation of the

Andreas Freier et al.

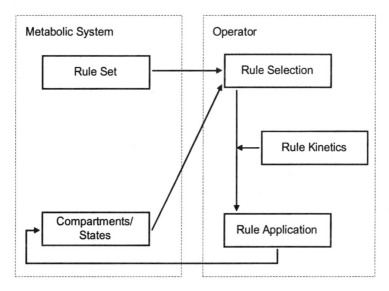

Figure 3.4 Mechanism of the derivation process in MetabSim operators.

enzyme. In MetabSim several kinetic behaviors are implemented (see figure 3.5): the constant flux and the linear, hyperbolic, and sigmoid dependencies from the substrate concentration. Their calculation requires the Michaelis-Menten constant (Km), the maximum reaction rate (V_{MAX}), and the enzyme type (allosteric, hyperbolic, etc.). The types read from the databases are mapped into the appropriate kinetic behavior and the calculation is done using the given parameters.

As an example, a rule network for the glycolysis pathway has been built up to simulate the consumption and the production of ATP. The following rules illustrate the stoichiometry of the metabolic system.

r_0 = ({D-glucose, ATP}, {glucose-6-phosphate, ADP}, {hexokinase, hyp})

r_1 = ({glucose-6-phosphate}, {(fructose-6-phosphate}, {phosphohexoseisomerase, hyp})

r_2 = ({fructose-6-phosphate, ATP}, {fructose-1,6-bisphosphate, ADP}, {phosphofructokinase, sgm})

r_3 = ({fructose-1,6-bisphosphate}, {glyceraldehyde-3-phosphate, dihydroxyacetone phosphate}, {aldolase, hyp})

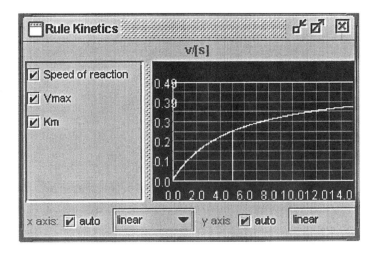

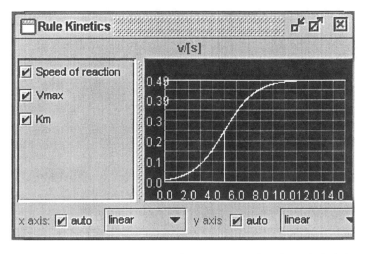

Figure 3.5 Examples of abstract rule kinetic types from hyperbolic behavior to sigmoid dependency function.

$r_4 = (\{\text{dihydroxyacetone phosphate}\}, \{\text{glyceraldehyde-3-phosphate}\},$
$\quad \{\text{triosephosphate isomerase, hyp}\})$

$r_5 = (\{\text{glyceraldehyde-3-phosphate, Pi}\}, \{\text{1,3-bisphosphoglycerate}\},$
$\quad \{\text{phosphoglyceraldehyde dehydrogenase, hyp}\})$

$r_6 = (\{\text{1,3-biphosphoglycerate, ADP}\}, \{\text{3-phosphoglycerate, ATP}\},$
$\quad \{\text{phosphoglycerate kinase, hyp}\})$

$r_7 = (\{\text{3-phosphoglycerate}\}, \{\text{2-phosphoglycerate}\},$
$\quad \{\text{phosphoglycerate mutase, hyp}\})$

$r_8 = (\{\text{2-phosphoglycerate}\}, \{\text{phosphoenolpyruvate}\}, \{\text{enolase, hyp}\})$

$r_9 = (\{\text{phosphoenolpyruvate, ADP}\}, \{\text{pyruvate, ATP}\},$
$\quad \{\text{pyruvatekinase, hyp}\})$

In this rule set, metabolites and enzymes are included. The metabolite D-glucose is consumed by the rules r_4 and r_5. One ATP is consumed in the first part of the glycolysis, and 2 ATP are produced in the second part. This concludes with the double application of rules r_7 and r_9. In figure 3.6 the system is drawn as a graph by the Metabuis environment. The application of the operator leads to the substance concentration development shown in figure 3.7.

MARGBench

The purpose of our project, which is supported by the German Research Council, is to present a virtual laboratory for the analysis of molecular processes (Hofestädt and Scholz, 1998). Therefore, we provide a full scalable system for a user-specific integration of different heterogeneous database systems and different interfaces for accessing the integrated data with special analysis tools (e.g., the simulation environment MetabSim, which was described earlier in this chapter).

Architecture of MARGBench

The architecture of the prototype is shown in figure 3.8 and is available on the Internet at http://www-bm.cs.uni-magdeburg.de/iti_bm/marg/. The system is divided into two main parts, the integration layer and the application layer. The integration layer consists

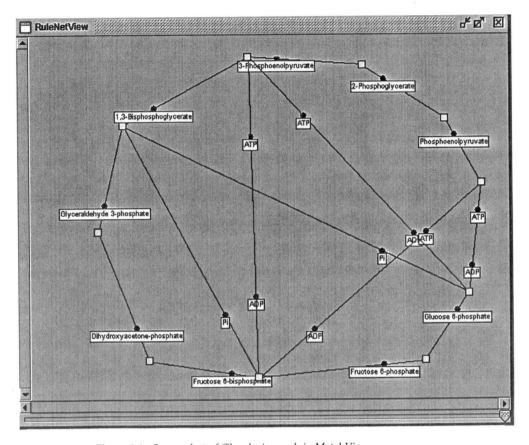

Figure 3.6 Screen shot of Glycolysis graph in MetabVis.

of three modules: data acquisition (BioDataServer), data storage (Bio-DataCache), and graphical data management (BioDataBrowser).

In general, the BioDataServer realizes a logical nonmaterialized database integration based on the concept of federated databases. A workable Internet access to the molecular biological databases is the main prerequisite for a database integration. Within this context several problems must be solved:

- Different interfaces (CGI, JDBC, etc.)
- Different query languages (SQL, OQL, nonstandardized, etc.)
- Different data structures (HTML, flat files, database objects, etc.)
- Different data models (ERM, OO, etc.)

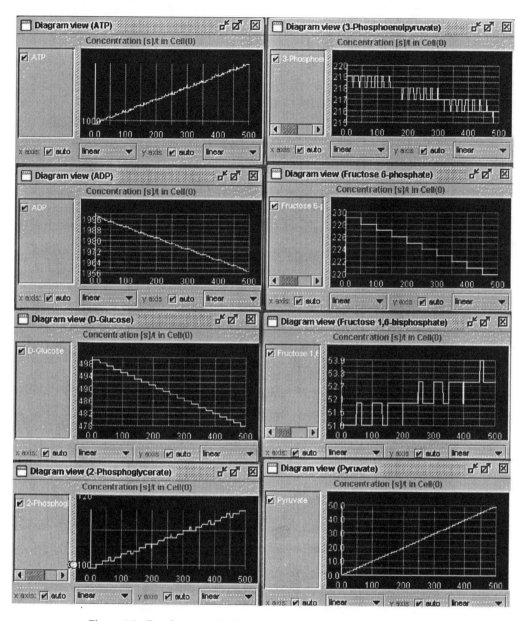

Figure 3.7 Development of substance concentration in MetabSim.

Information Fusion and Metabolic Network Control

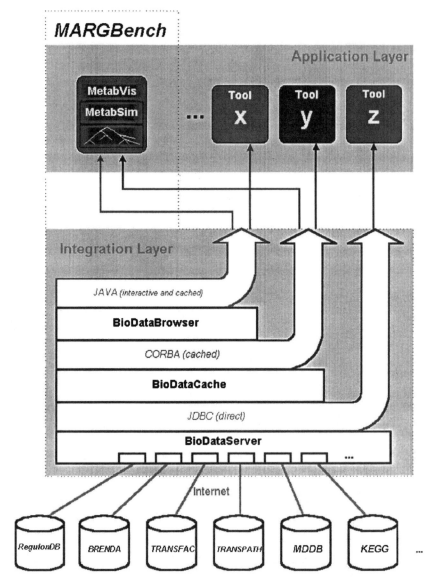

Figure 3.8 Architecture of MARGBench.

For such a homogeneous access to the several data sources, a functional interface was defined that is implemented by special software modules, called adapters. Using semiautomatic adapter generation, it will be possible to dynamically connect relevant data sources. In the case of an HTML data source, the adapter accesses the specific URL and parses the resulting HTML page. For this generation a description file is necessary. Such a description file contains structure and syntax information about the data source, which should be integrated into our system. This information enables a mapping between the data fields of the docked data sources and the attributes of an integrated user scheme.

Regarding the user requirements to the integrated data, it is possible to specify integrated data schemes. These schemes can be defined or manipulated, using a special language. They describe the accessible attributes of integrated data sources in transparent form.

In order to obtain a complete and wide spectrum of data, integration of as many databases as possible is recommended. Using this special integrated user scheme, the BioDataServer combines the outcomes of adapter queries into integrated global results, known as information fusion. The scheme is relational and defines the source for each attribute and its internal dependencies. This is the basis for performing logical data integration and provides a relational view of the fused data. Thus it is possible to retrieve integrated data by a subset of SQL queries.

As can be seen, an automatic mechanism is necessary to merge the partitioned data values from the various databases. This is one task of the BioDataServer, and can be solved using mathematical set operations.

The prerequisite to access the database integration server by computer programs is the definition of an interface. Because the server should be accessible via the Internet, a communication protocol and a query language must be specified. Many database systems support a subset of SQL as query language, which in turn is based on the relation model and is well standardized. This was the reason for supporting SQL elements by the BioDataServer. Different techniques in the field of interfaces for remote database access have been established (e.g., JDBC and ODBC). ODBC is supported only by Microsoft platforms. Therefore, the BioDataServer currently offers a JDBC driver, which provides a standardized database access to JAVA applications. Consequently, any JAVA platform can access the BioDataServer by related JAVA programs.

The main advantages of this BioDataServer are the transparent physical database access, the dynamic building of new nonmaterialized, logical integrated databases, a standardized access interface, a client-server capability, and platform independence. With the implemented JDBC driver the integration service of the BioDataServer is also independently useful for other external JAVA applications. A demonstration of the BioDataServer is available at `http://www-bm.cs.uni-magdeburg.de/BDSDemo/`.

The next level in the integration layer is the BioDataCache, which handles the local storage of the fused data in a user-specific integration database. Thus it is possible to build individual integrated databases that reflect the individual user's application requirements, and to perform data analysis, cleaning, improvements, and enrichment. The user is able to interactively specify and create the integration database in interface definition language (IDL) syntax. If the IDL is ready, the service modules will be generated automatically. The individual data import is based on specific, integrated user schemes of the BioDataServer, which must be defined previously. The access to these integration databases is possible by the Common Object Request Broker Architecture (CORBA) (OMG, 1991) and OQL (Cattell, 1994).

The BioDataCache provides the materialized layer for the data integration that is based on CORBA. The access for importing integrated data from the BioDataServer is possible using the JBDC driver of the integration layer. Furthermore, the BioDataCache has an integrated user scheme for the selection of attributes that should be integrated. An empty database related to the integrated scheme is generated automatically.

Once data from the integration service are read, they are stored in the underlying standard object-oriented database.

By storing the fused information in the cache, an separate new data source will be created. This new user-specific integration database represents a metadatabase and is comparable to a data warehouse. The offered CORBA interface, similar to the BioDataServer, enables other software tools to access the BioDataCache.

The third part in the integration layer is represented by the Bio-DataBrowser. It is automatically generated during the generation of an integration database. This module allows the user to graphically manage and browse the fused data. Its functions are similar to a Windows file browser. Furthermore, a JAVA interface is offered to access

Andreas Freier et al.

this component within JAVA applications. The development of DBMS-supported applications forces the programming of database-related user-interactive components to establish database connections, query the data, transmit the results, store the data, and so forth. The Bio-DataBrowser provides this feature and can be included as a component in different JAVA applications.

Application of MARGBench

The application of MARGBench is done at different levels, because the system consists of plugable components. Every user is able to access the components that are essential for his or her specific integration problem.

One possibility is the use of the online integration provided by the BioDataServer component. Here a uniform SQL interface with client/server and multiuser/multiclient feature is available. A case study for this kind of data analysis tool is available at `http://www-bm.cs.uni-magdeburg.de/phpMetaTool`.

The software METATOOL (Pfeiffer et al., 1999) is an independent system that has been written in the C programming language. Separate programs are necessary to generate the input file and to read the output files. The input for the integration architecture is obtained by the Bio-DataServer. This client program has been implemented using the PHP programming language, and enables the program to be used via the Internet.

The second way to use MARGBench is to access automatically generated integration databases. Once the integration database (BioData-Cache) has been established, the user can load his or her integration database with integrated data. For coupling application programs, the CORBA interface and the BioDataBrowser can be used.

As a reference application, MARGBench is used with the simulation tool MetabSim. It accesses a specially configured BioDataCache to obtain data and to build the rule network. First, a BioDataCache for storing information about metabolic networks is configured. After the data structures for the cache are established, the BioDataCache installation tool is used to compile and install an integration database (cache) called MetabNets. From the tool a CORBA interface is obtained for accessing the data structures previously defined. This interface can be used to submit database queries and to operate on cache objects by creating, modifying, and deleting operations. Data structures of MetabNets are, for instance, the classes metabolism, pathway, reaction, and enzyme. Once the cache scheme is defined, the user can start the integration procedure. During this process the BioDataServer is queried and objects of MetabNets are instantiated with the query results.

The data structures in the MetabSim program are also defined in CORBA. Now it is necessary to implement a mapping algorithm to produce rules out of the MetabNets objects. The program containing this algorithm accesses the BioDataBrowser and the BioDataCache. In this context, the browser is used for the interactive selection of metabolic pathways and single reactions. An example of enzyme information viewed with the BioDataBrowser is shown in figure 3.9. An active interface for the data exchange between BioDataBrowser and application is available. The application program (MetabSim) implements this data exchange interface. When the user selects one or more objects from the cache, the data exchange interface is called and the algorithm in MetabSim processes the transformation into MetabSim rules.

In order to handle large networks, all entities of the MetabSim simulation model are stored in a standard object-oriented database system. The MetabVis program is the front end of the simulation. With this graphical user interface, one can initiate the import of MetabNets data, modify the rule, run the simulation, and display the results. As an example of the result of access to different databases, figure 3.10 presents the gene regulation system of the CRP operon in *E. coli*.

DISCUSSION

The scientific world is at the beginning of the century of biotechnology. The progress of this new technology depends on the application of methods and concepts of computer science, because the exponential growth of experimental data must be handled. In other words, molecular database systems and analysis tools must be implemented. Apweiler et al. show in chapter 2 that more than 200 molecular database systems and hundreds of analysis tools are available via the Internet. For the analysis and synthesis of molecular processes, the integration of database systems is important. Therefore, the main goal of bioinformatics is to develop and implement the information technological infrastructure for molecular biology.

Various companies have already entered the market. The most important are Human Genome Sciences (USA), Celera Genomics (USA),

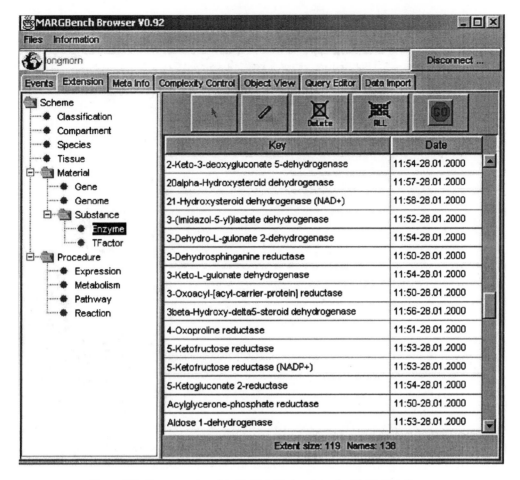

Figure 3.9 Enzymes stored in BioDataCache viewed with BioDataBrowser.

INCYTE Genomics (USA), Lion Bioscience (Germany), and Informax (USA). The backbone of such information systems is the integration of heterogeneous molecular database systems and analysis tools.

This chapter has described the current methods of database integration and the architecture of a molecular information system for the analysis and synthesis of gene-controlled metabolic networks. An integration shell was implemented that allows the semiautomatic implementation of a user-specific integration database which represents an information fusion process based on the integration of different molecular database systems. The implementation integrated, as a case study,

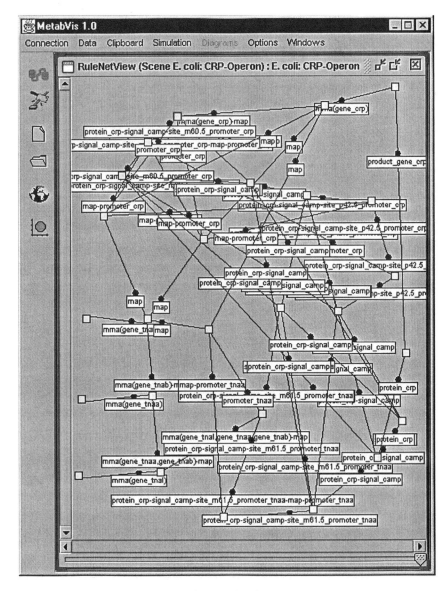

Figure 3.10 The CRP operon in *E. coli* viewed with MetabVis.

seven different molecular database systems and the rule-based simulation tool MetabSim, which allows the simulation of gene-controlled metabolic networks.

The rule-based method is easy to handle and also allows the abstract simulation of analytical effects. It does not have the vision of the virtual cell like Tomita (see chapter 11), because the theoretical results show that the complexity of the simulation of the complete metabolic processes is exponential. That means that only parts of the metabolism can be calculated. However, simulation is, and will be, one important point in understanding the function of gene-regulated metabolic pathways. Moreover, the algorithmic analysis of metabolic networks must be achieved. Chapters 10 and 7 show the state of the art of the algorithmic analysis of metabolic networks. To understand the logic of life, algorithms for the calculation of alignments of metabolic pathways or for the prediction of metabolic pathways based on rudimentary knowledge must be developed. Finally, the analysis process needs information systems that integrate analysis tools and simulation environments based on the integration of molecular database systems.

ACKNOWLEDGMENTS

This work was supported by the German Research Council and by the German Volkswagen Foundation.

REFERENCES

Baker, W., van den Broek, A., Camon, E., Hingamp, P., Sterk, P., Stoesser, G., and Tuli, M. A. (2000). The EMBL Nucleotide Sequence Database. *Nucleic Acids Res.* 28(1): 19–23.

Bork, P., and Bairoch, A. (1996). Go hunting in sequence databases but watch out for the traps. *Trends in Genetics* 12(10): 425–427.

Brutlag, D. L., Galpher, A. R., and Millis, D. H. (1991). Knowledge-based simulation of DNA metabolism: Prediction of enzyme action. *CABIOS* 7(1): 9–19.

Burks, C. (1999). Molecular biology database list. *Bioinformatics* 27(1): 1–9.

Cattell, R. G. (1994). *The Object Database Standard: ODMG-93.* San Mateo, Calif.: Morgan Kaufmann.

Codd, E. F. (1982). Relational database: A practical foundation for productivity. *Commun. ACM* 25(2): 109–113.

Conrad, S. (1997). Föderierte Datenbanksysteme: Konzepte der Datenintegration. *(Federated Database Systems: Concepts of Data Integration)*. Berlin and Heidelberg: Springer-Verlag.

Davidson, S. B., Overton, C., Tannen, V., and Wong, V. (1997). BioKleisli: A digital library for biomedical researchers. *Int. J. Digit. Libr.* 1: 36–53.

Etzold, T., Ulyanov, A., and Argos, P. (1996). SRS: Information retrieval system for molecular biology data banks. *Methods Enzymol.* 266: 114–128.

Franco, R., and Canela, E. (1984). Computer simulation of purine metabolism. *Eur. J. Biochem.* 144: 305–315.

Frishman, D., and Mewes, H. W. (1997). PEDANTic genome analysis. *Trends in Genetics* 13(10): 415–416.

Hofestädt, R. (1996). *Theorie der regelbasierten Modellierung des Zellstoffwechsels*. Aachen: Shaker.

Hofestädt, R., and Scholz, U. (1998). Information processing for the analysis of metabolic pathways and inborn errors. *BioSystems* 47(1–2): 91–102.

Hofestädt, R., and Thelen, S. (1998). Quantitative modeling of biochemical networks. *In Silico Biol.* 1(1): 39–53.

Kemp, G. J., Robertson, C. J., and Gray, P. M. (1999). Efficient access to biological databases using CORBA. *CCP11 Newsletter* 3.1(7); http://www.dl.ac.uk/CCP/CCP11/newsletter/vol3_1/.

Kohn, M. C., and Letzkus, W. (1982). A graph-theoretical analysis of metabolic regulation. *J. Theor. Biol.* 100: 293–304.

Luttgen, H., Rohdich, F., Herz, S., Wungsintaweekul, J., Hecht, S., Schuhr, C. A., Fellermeier, M., Sagner, S., Zenk, M. H., Bacher, A., and Eisenreich, W. (2000). Biosynthesis of terpenoids: YchB protein of *Escherichia coli* phosphorylates the 2-hydroxy group of 4-diphosphocytidyl-2C-methyl-D-erythritol. *Proc. Natl. Acad. Sci. USA* 97(3): 1062–1067.

Matsuda, H., Imai, T., Nakanishi, M., and Hashimoto, A. (1999). Querying molecular biology databases by integration using multiagents. *IEICE Trans. Inf. & Syst.* E82-D(1): 199–207.

Object Management Group (OMG). (1991). *The Common Object Request Broker: Architecture and specification*. OMG Document no 91.12.1. Framingham, Mass.: OMG.

Ogata, H., Goto, S., Sato, K., Fujibuchi, W., Bono, H., and Kanehisa, M. (1999). KEGG: Kyoto Encyclopedia of Genes and Genomes. *Nucleic Acids Res.* 27(1): 29–34.

Overbeek, R. (1992). Logic programming and genetic sequence analysis: A tutorial. In Krzysztof Apt (ed.), *Proceedings of Logic Programming*. Cambridge, Mass.: MIT Press.

Pfeiffer, T., Sánchez-Valdenebro, I., Nuño, J. C., Montero, F., and Schuster, S. (1999). METATOOL: For studying metabolic networks. *Bioinformatics* 15(3): 251–257.

Reddy, V. N., Mavrovouniotis, M. L., and Liebmann, M. N. (1993). Petri net representations in metabolic pathways. In L. Hunter, D. Searls, and J. Shavlik (eds.), *ISMB-93 Pro-*

ceedings: First International Conference on Intelligent Systems for Molecular Biology. Menlo Park, Calif.: AAAI Press; Cambridge, Mass.: MIT Press, pp. 328–336.

Senger, M., Glatting, K.-H., Ritter, O., and Suhai, S. (1995). X-Husar, an X-based graphical interface for the analysis of genomic sequences. *Comput. Methods Programs in Biomedicine* 46(2): 131–142.

Stoffers, H. J., Sonnhammer, E. L., Blommestijn, G. J., Raat, N. J., and Westerhoff, H. V. (1992). METASIM: Object-oriented modeling of cell regulation. *CABIOS* 8(5): 443–449.

Tomita, M., Hashimoto, K., Takahashi, K., Shimizu, T. S., Matsuzaki, Y., Miyoshi, F., Saito, K., Tanida, S., Yugi, K., Venter, J. C., and Hutchison, C. A. (1999). E-CELL: Software environment for whole-cell simulation. *Bioinformatics* 15(1): 72–84.

Topaloglou, T., Kosky, A., and Markowitz, V. (1999). Seamless integration of biological applications within a database framework. In T. Lengauer, R. Schneider, D. Boork, D. Brutlag, J. Glasgow, H.-W. Mewes, and R. Zimmer (eds.), *The Seventh International Conference on Intelligent Systems for Molecular Biology (ISMB '99), Heidelberg, Germany, August 6–10*. Menlo Park, Calif.: AAAI Press, pp. 272–281.

Waser, M., Garfinkel, L., Kohn, C., and Garfinkel, D. (1983). Computer modeling of muscle phosphofructokinase. *J. Theor. Biol.* 103: 295–312.

Xie, G., DeMarco, R., Blevis, R., and Wang, Y. (2000). Storing biological sequence databases in relational form. *Bioinformatics* 16(3): 288–289.

SUGGESTED READING

Bell, D., and Grimson, J. (1992). *Distributed Database Systems*. Reading, Mass.: Addison-Wesley.

Elmasri, R., and Navathe, S. B. (2000). *Fundamentals of Database Systems*. 3rd ed. Reading, Mass.: Addison-Wesley.

Middleton, J., Jones, M. L., and Pande, G. N. (1996). *Computer Methods in Biomechanics & Biomedical Engineering*. Amsterdam: Gordon and Breach.

Stephanopoulos, G. N., Aristidou, A. A., and Nielsen, J. (1998). *Metabolic Engineering: Principles and Methodologies*. London: Academic Press.

URLs FOR RELEVANT SITES

General Data Collections

Entrez. Entrez allows retrieval of molecular biology data and bibliographic citations from the NCBI's integrated databases. `http://www.ncbi.nlm.nih.gov`

SRS. The Sequence Retrieval System (SRS) is a huge, systematic collection of molecular databases and analysis tools that are uniquely formatted and accessible. `http://srs6.ebi.ac.uk`

Genes

CEPH. This is a database of genotypes for all genetic markers that have been typed in the CEPH and for reference families for linkage mapping of human chromosomes. http://www.cephb.fr

EMBL. The EMBL Nucleotide Sequence Database is a framed compilation of all known DNA and RNA sequences. http://www.ebi.ac.uk

GDB. The GDB is the offical central database for all information that is collected in the Human Genome Project. http://gdbwww.gdb.org/

GenBank. GenBank is the NIH genetic sequence database. It is an annotated collection of all publicly available DNA sequences. http://www.ncbi.nlm.nih.gov

GeneCards. This database covers data of human genes, their products, and their involvement in diseases. http://bioinfo.weizmann.ac.il/cards/

HGMD. The Human Gene Mutation Database represents an attempt to collate known (published) gene lesions responsible for human inherited disease. http://www.uwcm.ac.uk

MGD. The Mouse Genome Database contains information on mouse genetic markers, molecular segments, phenotypes, comparative mapping data, etc. http://www.informatics.jax.org

MKMD. Mouse Knockout and Mutation Database is a regularly updated database of mouse genetic knockouts and mutations. References are directly linked to Evaluated Medline. Published by *Current Biology*. http://www.biomednet.com

MTIR. Data about the expression of muscle-specific genes are available in this database. http://agave.humgen.upenn.edu

PAHdb. Database that provides access to up-to-date information about mutations at the phenylalanine hydroxylase locus, including additional data. http://blizzard.cc.mcgill.ca

SCPD. This is a specialized promoter database of *Saccharomyces cerevisiae*. http://cgsigma.cshl.org/jian

VBASE. VBASE is a comprehensive directory of all human germ line variable region sequences compiled from over a thousand published sequences. http://www.mrc-cpe.cam.ac.uk

Proteins and Enzymes

CATH. CATH is a hierarchical classification of protein domains. A lexicon that includes text and pictures describing protein class and architecture is available. http://www.biochem.ucl.ac.uk

ENZYME. ENZYME is a repository of information relative to the nomenclature of enzymes. http://www.expasy.ch

LIGAND. The Ligand Chemical Database for Enzyme Reactions is designed to provide the linkage between chemical and biological aspects of life in light of enzymatic reactions. `http://www.genome.ad.jp`

PDB. PDB is the single international repository for the processing and distribution of 3-D macromolecular structure data primarily determined experimentally by X-ray crystallography and NMR. `http://www.pdb.bnl.gov/`

PIR. This database is a comprehensive, annotated, and nonredundant set of protein sequence databases in which entries are classified into family groups and alignments of each group are available. `http://pir.georgetown.edu`

PMD. The Protein Mutant Database will be valuable as a basis of protein engineering. It is based on literature (not on proteins); that is, each entry corresponds to an article that describes protein mutations. `http://www.genome.ad.jp`

PRF. The Peptide Institute, Protein Research Foundation, collects information related to amino acids, peptides, and proteins: articles from scientific journals, peptide/protein sequence data, data on synthetic compounds, and molecular aspects of proteins. `http://www.prf.or.jp`

PRINTS. PRINTS is a compendium of protein fingerprints. A fingerprint is a group of conserved motives used to characterize a protein family. `http://www.biochem.ucl.ac.uk`

PROSITE. PROSITE is a database of protein families and domains. It consists of biologically significant sites, patterns, and profiles that help to reliably identify to which known protein family (if any) a new sequence belongs. `http://www.expasy.ch/prosite/`

REBASE. REBASE, the Restriction Enzyme data BASE is a collection of information about restriction enzymes (methylases), the microorganisms from which they have been isolated, recognition sequences, cleavage sites, etc. `http://rebase.neb.com`

SWISS-PROT. SWISS-PROT is a curated protein sequence database that strives to provide a high level of annotations, a minimal level of redundancy, and a high level of integration with other databases. `http://expasy.hcuge.ch/sprot/sprot_top.html`

Pathways

CSNDB. The Cell Signaling Networks DataBase is a data and knowledge base for signaling pathways of human cells. `http://geo.nihs.go.jp`

ExPASy. This database contains links to the ENZYME Database and, for each entry, also contains links to all maps of Boehringer Biochemical Pathways in which this entry appears. `http://www.expasy.ch`

KEGG. The Kyoto Encyclopedia of Genes and Genomes is an effort to computerize current knowledge of molecular and cellular biology in terms of the information pathways that consist of interacting molecules or genes. `http://www.genome.ad.jp/kegg`

WIT. WIT is a web-based system to support the curation of function assignments made to genes and the development of metabolic models. `http://wit.mcs.anl.gov/WIT2`

Gene Regulation

EPD. The Eukaryotic Promoter Database is a collection of eukaryotic promoters in form of DNA sequences. `ftp://ftp.ebi.ac.uk/pub/databases/epd/`

GRBase. The Growth Regulation Database contains information about proteins, genes, and sequences in the field of gene regulation. `http://www.access.digex.net`

RegulonDB. This is a database on transcriptional regulation in *E. coli.* `http://www.cifn.unam.mx.unam.mx/Computational_Genomics/regulondb`

TRANSFAC. This database compiles data about gene regulatory DNA sequences and protein factors binding to and acting through them. `http://transfac.gbf.de`

TRRD. The Transcription Regulatory Regions Database is a curated database designed for accumulation of experimental data on extended regulatory regions of eukaryotic genes. `http://www.bionet.nsc.ru/SRCG/index.html`

Metabolic Diseases

BIOMDB. This consists of collected data on mutations causing tetrahydrobiopterin deficiencies. `http://www.unizh.ch`

NORD. The NORD Resource Guide is an invaluable resource for families, health care professionals, and libraries. `http://www.stepstn.com`

OMIM. The OMIM (Online Mendelian Inheritance in Man) database is a catalog of human genes and genetic disorders written and edited by Dr. Victor A. McKusick and his colleagues. `http://www3.ncbi.nlm.nih.gov`

PATHWAY. PATHWAY, a database of inherited metabolic diseases, is divided into two sections: substances and diseases. `http://oxmedinfo.jr2.ox.ac.uk`

PEDBASE. PEDBASE is a database of pediatric disorders. Entries are listed alphabetically by disease or condition name. `http://www.gretmar.com`

RDB. The Rare Disease Database is a system for delivery of understandable medical information to the public, including patients, families, physicians, medical institutions, and support organizations. `http://www.rarediseases.org`

II   Gene Regulation: From Sequence to Networks

4 Specificity of Protein-DNA Interactions

Gary D. Stormo

Gene expression is often regulated transcriptionally through the action of protein factors that bind to DNA and affect the rate of initiation of the (usually) nearby promoter, either increasing it (activators) or decreasing it (repressors). In order for this process to be promoter-specific, rather than affecting the expression of all genes, at least some of the transcription factors must recognize and bind to specific DNA sequences. The purpose of this article is to define a quantitative measure of specificity and describe various models to represent the specificity of a particular protein that can be used to predict binding sites in genomic DNA. It also briefly describes how those models can be used in pattern recognition methods to identify regulatory sites from sets of coregulated genes.

QUANTITATIVE SPECIFICITY

Consider a DNA-binding protein that binds to sites that are l-long. There are 4^l such sequences, and we refer to each one as S_i with $1 \leq i \leq 4^l$ (i and l are integers). Each sequence binds to the protein with some free energy, denoted by H_i. Of course that will depend on the binding conditions, but for simplicity it is assumed there is some standard condition that is always used. (It is very interesting to study changes in H_i as a function of the reaction conditions, but that issue is not addressed in this article.) Figure 4.1a represents the list of all 4-long sequences and the binding energy to each one for some hypothetical protein. The energies are shown relative to a sequence with average affinity, so that those with negative values are preferred and those with positive values are discriminated against by the protein. In general we

a)

AAAA	2.63
AAAC	2.63
AAAG	2.63
AAAT	0.46
AACA	1.79
AACC	1.79
AACG	1.79
AACT	−0.38
⋮	⋮
AGCG	−0.25
AGCT	−2.42
AGGA	0.59
⋮	⋮
CTTT	−0.37
GAAA	4.01
⋮	⋮
TTTG	2.63
TTTT	0.46

b)

	1	2	3	4
A	−0.55	+1.38	+0.42	+1.38
C	0	+0.55	−0.42	+1.38
G	+0.83	−0.66	+0.42	+1.38
T	+0.83	+0.42	0	−0.79

Figure 4.1 Specificity of a hypothetical DNA-binding protein. (a) The complete list of binding free energies to all 256 4-long sequences. In this example the units are kcal/mole, and the binding energies are relative to the site with average affinity. (b) The weight matrix model that provides the same binding energies for all sequences. The energy for an individual sequence is the sum of the values for the bases that correspond to that sequence in each position (see figure 4.2).

care only about the differences in binding energies, because those determine the probabilities of binding to different sequences. For example, the ratio of binding affinities, K_i, K_j, for two sequences, S_i, S_j, is just

$$\frac{K_i}{K_j} = e^{-H_i + H_j} \tag{4.1}$$

At equilibrium the distribution of sites bound by the protein, f_i, can be obtained from the list of binding energies, H_i, and the distribution of potential binding sites, g_i, from the Boltzmann equation (Heumann et al., 1994):

$$f_i = \frac{g_i e^{-H_i}}{\sum_{j=1}^{4^l} g_j e^{-H_j}} = \frac{g_i e^{-H_i}}{Z}. \tag{4.2}$$

Z is the partition function and assures that $\sum_i f_i = 1$. The temperature is not included because it is assumed to be one of the conditions that is held constant. g_i is the number of occurrences of each sequence, S_i, in the set of possible binding sites. In vivo, g_i would be the composition of the genome in words of length l, or at least the composition of sites that are available for binding to the protein. As stated above, it is only the difference in binding energies that matters, because f_i is unaffected by replacing all values of H_i by $H_i + c$.

One common choice for the baseline of energy is to set the energy for the preferred sequence to 0, so that all other sequences have positive energy (Berg and von Hippel, 1987). We often choose the average affinity as the baseline for the energy, such that $Z = G$, the total number of possible binding sites, which is the total number of available sites in the genome when considering the in vivo situation. We can substitute $p_i = g_i/G$, where p_i represents the probability of each sequence S_i in the set of possible binding sites.

Rearranging equation 4.2 provides a conceptually simple, but technically difficult, method to determine the binding energies of each sequence:

$$H_i = -\ln\frac{f_i}{p_i} \tag{4.3}$$

That is, you can imagine mixing all possible sequences together in known amounts, p_i, and then measuring the fraction of each sequence bound to the protein at equilibrium, f_i. The logarithms of those ratios give exactly the energy values desired. Furthermore, at equilibrium the average binding energy is simply

$$\langle H_i \rangle = -\sum_i f_i \ln\frac{f_i}{p_i}, \tag{4.4}$$

which is the relative entropy, or Kullbach-Leibler distance, between the two distributions of the potential binding sites and the bound sites. The derivation of this simple, but useful, relationship has been achieved through the use of the Boltzmann equation, but there is an equivalent derivation using Bayes's theorem for conditional probabilities (Heumann et al., 1994).

MODELS FOR SPECIFICITY

While in theory we could measure H_i for every sequence and simply put the measurement into a lookup table to use whenever needed, as in figure 4.1(a), the task is impractical for typical sizes of l. Instead, one usually employs some sort of model that provides an estimate of H_i for all sequences. The simplest model is to assume that the protein binds to some set of sequences (perhaps only one) and not at all (or at least very much worse) to any other sequence. If it binds equally well to all of its binding sites, each site gets some energy x, and every other sequence has energy $+\infty$ (or some large increase over the energy of the sites). For example, the specificity shown in figure 4.1a could be modeled as a negative free energy for the preferred sequence, AGCT, and a large positive value for all other sequences, using the assumption that only AGCT shows significant binding to the protein. Such an assumption might be valid for some proteins, such as restriction and modification enzymes that are extremely sequence-specific in their activities. However, most regulatory proteins are much less specific and show significant binding to sequences other than their preferred sites. The model could be modified to allow for binding to sequences similar to AGCT, perhaps those sequences with one mismatch (there are 12 such sequences). Any method that searches for binding sites using a consensus sequence, even allowing for some number of mismatches, implicitly assumes such a model for binding energies.

The next, more complicated model employs a "weight matrix" $H(b, m)$ that contains an energy for each base, $b \in \{A, C, G, T\}$, at each position, $1 \le m \le l$, in the site (Stormo, 1990, 2000). Figure 4.1b shows such a model for the hypothetical protein. The binding energy to any sequence is $H_i = H(b, m) \cdot S_i$. In this notation S_i is a matrix with 1 for the base that occurs at each position and 0 elsewhere, so the dot product selects the energies that correspond to the sequence, as shown in figure 4.2. The main advantage of weight matrix methods over consensus sequence methods is that they allow different mismatches from the preferred sequence to have different effects on the binding site predictions. For the protein in figure 4.1, mismatches to the T at position 4 would have dramatic effects on binding, whereas all other mismatches would have smaller effects. In fact, many double mismatches and even the triply mismatched sequence CTTT bind with higher affinity than a single

Gary Stormo

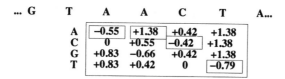

Figure 4.2   The energy of any particular sequence is determined by summing the matrix values that correspond to the sequence. The boxed values correspond to the sequence AACT and give a total binding energy of −0.38 kcal/mole. As the matrix is moved along, the sequence different values will be added together to give the energy of the corresponding sequence.

mismatch at position 4. Such differential effects are taken into account naturally by the weight matrix model, and they are not uncommon in the specificity of regulatory proteins, where some positions may be highly conserved and others highly variable.

The weight matrix model assumes that the total binding energy is the sum of the interactions at each position. For the hypothetical protein of figure 4.1 we assume complete additivity, so that all 256 values in figure 4.1a can be obtained precisely from the 16 parameters of the model in figure 4.1b. That additivity assumption may be reasonable for some proteins, but may not be good enough for others. The model can be made more complicated with a matrix that has a row for each dinucleotide rather than for each base (and only $l - 1$ columns). That would accommodate nonindependent interactions at adjacent positions in the binding site. If that is still not good enough, the matrix can be made with trinucleotide rows, or higher. (Of course this can be done as a Markov chain rather than an explicit matrix of all possible base combinations, but that only changes the notation.)

At some point the weight matrix method must provide a perfect "model" for the interaction because it becomes a matrix with 4^l rows and only one column, the lookup table of figure 4.1a. For practical reasons, primarily limited data, we usually assume that a simple weight matrix (bases by positions) will be a sufficiently good model, but we have to allow the data to drive us to more complicated models when necessary. Exactly when a particular model is "good enough" often depends on how it will be used.

Given that we want to represent the specificity of a protein with a weight matrix, there are several approaches that might be used to obtain one. Since the weight matrix has only $4l$ parameters (and since we care only about the differences in each column, just $3l$ degrees of freedom), one might simply measure the change in binding energy to all single base changes from the preferred sequence (Sarai and Takeda, 1989; Takeda et al., 1989; Fields et al., 1997). Alternatively, given a collection of sequences with known binding energy, one could find the matrix that provides the best fit to that quantitative data (Stormo et al., 1986; Barrick et al., 1994). This latter approach has the advantage of indicating whether the weight matrix provides a sufficiently good model of the interaction. If not, even the best fit will be fairly poor and more complicated models will be necessary (Stormo et al., 1986).

Probably the easiest and most common method is to collect a set of known binding sites and develop the matrix from a statistical analysis of those sites. The known sites may be naturally occurring or they may be selected from a random pool (Schneider et al., 1986; Fields et al., 1997). Because we assume a binding energy model of the form $H(b, m)$, we can treat the positions independently and convert the aligned collection of known sites into a frequency matrix $F(b, m)$ that counts the fraction of each base at each position in the aligned sites, as in figure 4.3. If we make the further assumption that the probability of a sequence, p_i, is determined by its composition, and each base b has probability $P(b)$, then the best estimate of the energy matrix is

$$H(b, m) = -\ln \frac{F(b, m)}{P(b)}. \tag{4.5}$$

This method of estimating the binding energy for a protein based on a collection of known sequences originated with Staden (1984), except that he set $P(b) = 1$. Schneider (1997) uses the same form with $P(b) = 0.25$. Berg and von Hippel (1987) substituted $\max_b F(b, m)$ for $P(b)$ at each position m, so that the most common base is assigned energy 0 and all other bases get positive energy.

All of these approaches are equivalent if the $P(b)$ are all approximately equal, but can be inappropriate if they are not or if the p_i are not well approximated by their composition. It is easy to show that if the p_i are well approximated by the composition, then the formula of equation 4.5 provides the values of $H(b, m)$ that maximize the probability of binding to the known sites (Heumann et al., 1994). Then the average binding energy to the known sites at equilibrium is

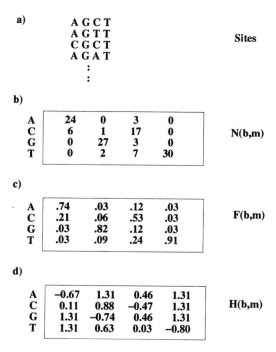

a)

```
A G C T
A G T T        Sites
C G C T
A G A T
    :
    :
```

b)

A	24	0	3	0
C	6	1	17	0
G	0	27	3	0
T	0	2	7	30

N(b,m)

c)

A	.74	.03	.12	.03
C	.21	.06	.53	.03
G	.03	.82	.12	.03
T	.03	.09	.24	.91

F(b,m)

d)

A	−0.67	1.31	0.46	1.31
C	0.11	0.88	−0.47	1.31
G	1.31	−0.74	0.46	1.31
T	1.31	0.63	0.03	−0.80

H(b,m)

Figure 4.3 Determining the matrix for a set of binding sites. (a) An example set of binding sites. These are taken from the list of possible binding sites, based on their probabilities. (b) The summary matrix of the 30 aligned sites, partially listed in (a). (c) The frequency matrix, using a small sample size correction of adding 1 to each value in the summary matrix, and dividing by the total (34). (d) The resulting energy matrix, using $P(b) = 0.25$ for all bases. These values have been converted back into units of kcal/mole, using $RT = 0.62$.

$$\langle H_i \rangle_{sites} = -\sum_b \sum_m F(b, m) \ln \frac{F(b, m)}{P(b)}, \qquad (4.6)$$

the relative entropy between the base frequencies at the sites and the genomic base frequencies (Schneider et al., 1986; Stormo and Fields, 1998). This is a very useful objective function for the pattern recognition methods we employ because the set of sites that maximize the relative entropy would be those sites with the highest probability of binding to the protein under the assumptions of equilibrium, additivity, and a random background. Some of those assumptions can be removed without increasing the complexity of the problem too much, and we do so when it seems useful.

DISCOVERING BINDING SITES

The previous section described how we can represent the specificity of a DNA-binding protein as a weight (or energy) matrix, and how we can obtain the best estimate for that matrix, given a collection of known sites. However, often the problem of interest is how to discover the binding site pattern (energy matrix) for a transcription factor, given only a collection of genes that are known to be regulated by it. For example, an expression array experiment can indicate sets of genes that apparently are coordinately regulated (Spellman et al., 1998), and therefore may each have a binding site for a common transcription factor. We don't know the energy matrix for the protein, or where the binding sites are, except that they are most likely to be in the promoter regions for the genes. The goal then is to find an energy matrix such that each of the coregulated sequences has at least one high-probability binding site.

Several approaches now exist to find such sites. There are word-based methods that try to find overrepresented words in promoter regions. This approach was first presented by Galas et al. (1985), has been modified several times, and most recently has been used effectively on sets of coregulated yeast genes (van Helden et al., 1998; Brazma et al., 1998). Although the algorithms don't work this way, it is possible to think of them as dividing all potential binding sites into two classes, those that bind (with some energy x) and those that don't (energy $+\infty$). Their objective is to maximize the probability of binding to the coregulated promoter regions, which will increase if they have more of the words that bind to the protein.

These methods can take the background probabilities, p_i, into account by comparing the frequencies of the words identified to their frequencies in other promoters (van Helden et al., 1998). These methods don't really produce an accurate binding energy description, because most proteins do not bind with "all or nothing" affinities. But they can still be effective at identifying the regulatory sites, which can then be analyzed in other ways. This is especially true if the binding sites are relatively short and highly conserved. However, in general, matrix-based methods tend to be more reliable because they can also identify the short, highly conserved sites and can allow for more variability in the sites.

There are four general methods, and several specific programs, for identifying sites via weight matrix approaches: a greedy algorithm, expectation-maximization (EM), a Gibbs's sampling method, and a network method for maximizing specificity. The goal in each case is the same, so they often return the same solution. However, all of the methods are heuristic and may not find the true maximum of their objective. Each algorithm attempts to maximize the objective in a different way, and so may be susceptible to different local optima. It is usually advisable to try multiple methods to see if consistent results are obtained. Furthermore, there is usually a variety of parameters that can be set which affect how the programs run, and exploring various parameter values may be worthwhile.

Our CONSENSUS program is a greedy algorithm that builds up the total alignment of sites by adding new ones at each iteration (Stormo and Hartzell, 1989; Hertz et al., 1990; Hertz and Stormo, 1999). It is somewhat similar to CLUSTAL (Higgins et al., 1996) in that it develops the full multiple alignment of sites progressively, but has several important differences. First, it is specifically searching for ungapped local alignments of length l. (The version WCONSENSUS does not require the length to be specified, and will search for the optimum length during the run.) Second, the program does not rely on a single best alignment at any step, but at each iteration keeps many (a user-defined parameter, typically set to 1000 or more) potential multiple alignments to be compared against the remaining sequences.

Third, the alignments are ranked by either their information content or their p-values. Originally we ranked alignments based on their information content because of its relationship to the average binding energy of the predicted sites. However, information content is always nonnegative and, by itself, not directly related to the significance of an alignment. But by taking into account the length and composition of the promoter regions, as well as some other relevant parameters, it is possible to compute the probability of finding by chance an alignment with an information content above some value, which gives us the p-value for each alignment (Hertz and Stormo, 1999). Ranking by p-value allows direct comparison of the significance of alignments of different lengths, different number of sites included, and even determination of whether a particular pattern is symmetric or not. This last issue, whether a binding site is a symmetric pattern, can indicate whether a regulatory protein binds as a homodimer, as is fairly common. Given an alignment of sites that are intrinsically symmetric, it is always possible to obtain an asymmetric alignment with greater (or equal) information content, because each individual site has two ways to be aligned. Only by considering the reduced number of degrees of freedom imposed by the symmetry of the pattern is its symmetry apparent. Thus use of the p-value can provide improved estimates of the optimal binding energy parameters.

The EM algorithm and the Gibbs's sampling method are similar overall, but have one important difference (Lawrence and Reilly, 1990; Lawrence et al., 1993). Both are iterative algorithms that alternate between the two steps shown in figures 4.2 and 4.3. They usually start with an arbitrary alignment of sites, but can also start with an arbitrary matrix. Given a set of sites, a matrix is produced using the method in figure 4.3. Then, using the matrix, the probability of the protein binding to every site is determined from its predicted energy, as in figure 4.2. This procedure is iterated and is guaranteed to converge in the case of EM, and tends to higher values of information content with Gibbs's sampling.

The main difference is in how the sites from which the matrix is made are chosen. In EM, every site contributes to the alignment but is weighted by its probability. In Gibbs's sampling, only one site is selected in one sequence, using the matrix derived from the predicted sites in the other sequences. The site chosen is based on its probability; the higher the probability, the more likely it is to be chosen, but any site may be chosen at each iteration. This stochastic aspect of the Gibbs's sampling procedure makes it able to escape from local optima, and therefore it is more robust than EM. The objective used in the Gibbs's sampling method is identical to information content, using a small size correction, as in figure 4.3 (see equations 1 and 2 in Lawrence et al., 1993). The EM algorithm has been implemented in the MEME package of programs (Bailey and Elkan, 1994; Grundy et al., 1996). Several newer versions of Gibbs's sampling algorithm exist (Spellman et al., 1998; Roth et al., 1998; Wasserman et al., 2000).

Each of the weight matrix methods described thus far considers only the promoters from the coregulated genes, although potential binding sites from the rest of the genome influence the probability of those sites being bound (equation 4.2). They assume that the rest of the genome can be modeled as a random sequence with a given composition, $P(b)$.

Gary Stormo

However, for many sets of coregulated promoters this assumption can be misleading. For example, yeast promoters often contain strings of A/T-rich sequences that occur much more often than expected, even given the A/T-rich genome. Such patterns clearly are statistically significant, given the random genome assumption, but they are not specific to the set of coregulated genes. Rather, they are patterns that are common in many promoters, and so cannot be the sought-after binding sites responsible for the coregulation.

To address this issue we developed an alternative algorithm, implemented in the program ANN-Spec (Heumann et al., 1994; Workman and Stormo, 2000). It still uses a weight matrix model for the binding energy, but it determines the partition function explicitly using the entire genome, or at least all of the promoter regions for eukaryotic organisms. It starts with an arbitrary matrix and then selects binding sites from each sequence using Gibbs's sampling approach. It also calculates the partition function and then determines the gradient of the site probabilities as a function of the weight matrix parameters. It follows this gradient until convergence.

This method is still heuristic, but tends to work as well as the other methods we've tried on both simulated sequences (Workman and Stormo, 2000) and real promoters from yeast and other eukaryotes (unpublished results). When the background is approximately random, its objective function is the same as information content (Heumann et al., 1994), so it should give the same results as the other methods. But its advantage is that it specifically identifies patterns that are able to distinguish the coregulated genes from other promoters, and therefore does not find the A/T-rich patterns that the other methods might (Workman and Stormo, 2000).

Of course there are other methods to avoid those common patterns in postprocessing steps. The AlignACE version of Gibbs's sampling algorithm identifies many potentially significant patterns and then filters out the uninteresting ones based on a variety of criteria, including the frequencies of the pattern in the coregulated genes and the rest of the genome (Hughes et al., 2000).

FUTURE DIRECTIONS

One of the limitations of the approaches described is that they tend to view proteins binding independently of one another. It is easy to accommodate situations in which two or more proteins compete for binding to mutually exclusive sites. More problematic are situations where proteins bind cooperatively. It is known that eukaryotic promoters are not usually regulated by the binding of a single transcription factor, as often happens in prokaryotes. Rather, it is common for multiple factors binding to multiple sites to be involved in regulating transcription. Provided each protein shows sufficient specificity on its own to be identified, the methods already described should be able to find all of the binding sites, either simultaneously or through iterated application of the program. But if the patterns themselves are significant only through their joint occurrence, then alternative methods are necessary that can specifically look for composite patterns that are significant jointly, even if not individually. Once combinations of individual patterns that lead to specific regulatory responses are identified, then the combinatorial network of gene regulation can be more fully modeled.

There has been some work in the area of defining complete, multicomponent regulatory patterns (Frech et al., 1997; Hu et al., 2000; Liu et al., 2001; GuhaThakurta and Stormo, 2001), but more work is needed. Additional extensions would include knowledge of operon organization and regulatory proteins, as well as transcriptome experiments.

ACKNOWLEDGMENT

The work described from my lab has been supported by NIH grant HG00249.

REFERENCES

Bailey, T. L., and Elkan, C. P. (1994). Fitting a mixture model by expectation maximization to discover motifs in biopolymers. *Proc. Int. Conf. Intell. Sys. Mol. Biol.* 2: 28–36.

Barrick, D., Villanueba, K., Childs, J., Kalil, R., Schneider, T. D., Lawrence, C. E., Gold, L., and Stormo, G. D. (1994). Quantitative analysis of ribosome binding sites in *E. coli. Nucleic Acids Res.* 22: 1287–1295.

Berg, O. G., and von Hippel, P. H. (1987). Selection of DNA binding sites by regulatory proteins. Statistical-mechanical theory and application to operators and promoters. *J. Mol. Biol.* 193: 723–750.

Brazma, A., Jonassen, I., Vilo, J., and Ukkonen, E. (1998). Predicting gene regulatory elements *in silico* on a genomic scale. *Genome Res.* 8: 1202–1215.

Fields, D. S., Yi-yuan, H., Al-Uzri, A., and Stormo, G. D. (1997). Specificity of the Mnt repressor determined by quantitative DNA sequencing. *J. Mol. Biol.* 271: 178–194.

Frech, K., Danescu-Mayer, J., and Werner, T. (1997). A novel method to develop highly specific models for regulatory units detects a new LTR in GenBank which contains a functional promoter. *J. Mol. Biol.* 270: 674–687.

Galas, D. J., Eggert, M., and Waterman, M. S. (1985). Rigorous pattern-recognition methods for DNA sequences. Analysis of promoter sequences from *Escherichia coli*. *J. Mol. Biol.* 186: 117–128.

Grundy, W. N., Bailey, T. L., and Elkan, C. P. (1996). ParaMEME: A parallel implementation and a web interface for a DNA and protein motif discovery tool. *Comput. Appl. Biosci.* 12: 303–310.

GuhaThakurta, D., and Stormo, G. D. (2001). Identifying target sites for cooperatively binding factors. *Bioinformatics* 17: 608–621.

Hertz, G. Z., Hartzell, G. W. III, and Stormo, G. D. (1990). Identification of consensus patterns in unaligned DNA sequences known to be functionally related. *Comput. Appl. Biosci.* 6: 81–92.

Hertz, G. Z., and Stormo, G. D. (1999). Identifying DNA and protein patterns with statistically significant alignments of multiple sequences. *Bioinformatics* 15: 563–577.

Heumann, J. M., Lapedes, A. S., and Stormo, G. D. (1994). Neural networks for determining protein specificity and multiple alignment of binding sites. *Intell. Sys. Mol. Biol.* 2: 188–194.

Higgins, D. G., Thompson, J. D., and Gibson, T. J. (1996). Using CLUSTAL for multiple sequence alignments. *Methods Enzymol.* 266: 383–402.

Hu, Y.-J., Sandmeyer, S., McLaughlin, C., and Kibler, D. (2000). Combinatorial motif analysis and hypothesis generation on a genomic scale. *Bioinformatics* 16: 222–232.

Hughes, J. D., Estep, P. W., Tavazoie, S., and Church, G. M. (2000). Computational identification of cis-regulatory elements associated with groups of functionally related genes in *Saccharomyces cerevisiae*. *J. Mol. Biol.* 296: 1205–1214.

Lawrence, C. E., Altschul, S. F., Boguski, M. S., Liu, J. S., Neuwald, A. F., and Wootton, J. C. (1993). Detecting subtle sequence signals: A Gibbs sampling strategy for multiple alignment. *Science* 262: 208–214.

Lawrence, C. E., and Reilly, A. A. (1990). An expectation maximization (EM) algorithm for the identification and characterization of common sites in unaligned biopolymer sequences. *Proteins* 7: 41–51.

Liu, X., Brutlag, D. L., and Liu, J. S. (2001). Bioprospector: Discovering conserved DNA motifs in upstream regulatory regions of co-expressed genes. *Pac. Symp. Biocomput.* 6: 127–138.

Roth, F. P., Hughes, J. D., Estep, P. W., and Church, G. M. (1998). Finding DNA regulatory motifs within unaligned noncoding sequences clustered by whole-genome mRNA quantitation. *Nat. Biotechnol.* 16: 939–945.

Sarai, A., and Takeda, Y. (1989). Lambda repressor recognizes the approximately 2-fold symmetric half-operator sequences asymmetrically. *Proc. Natl. Acad. Sci. USA* 86: 6513–6517.

Schneider, T. D. (1997). Information content of individual genetic sequences. *J. Theor. Biol.* 189: 427–441.

Schneider, T. D., Stormo, G. D., Gold, L., and Ehrenfeucht, A. (1986). Information content of binding sites on nucleotide sequences. *J. Mol. Biol.* 188: 415–431.

Spellman, P. T., Sherlock, G., Zhang, M. Q., Iyer, V. R., Anders, K., Eisen, M. B., Brown, P. O., Botstein, D., and Futcher, B. (1998). Comprehensive identification of cell cycle-regulated genes of the *yeast Saccharomyces cerevisiae* by microarray hybridization. *Mol. Biol. Cell* 9: 3273–3297.

Staden, R. (1984). Computer methods to locate signals in nucleic acid sequences. *Nucleic Acids Res.* 12: 505–519.

Stormo, G. D. (1990). Consensus patterns in DNA. *Methods Enzymol.* 183: 211–221.

Stormo, G. D. (2000). DNA binding sites: Representation and discovery. *Bioinformatics* 16: 16–23.

Stormo, G. D., and Fields, D. S. (1998). Specificity, free energy and information content in protein-DNA interactions. *Trends Biochem. Sci.* 23: 109–113.

Stormo, G. D., and Hartzell, G. W. III. (1989). Identifying protein-binding sites from unaligned DNA fragments. *Proc. Natl. Acad. Sci. USA* 86: 1183–1187.

Stormo, G. D., Schneider, T. D., and Gold, L. (1986). Quantitative analysis of the relationship between nucleotide sequence and functional activity. *Nucleic Acids Res.* 14: 6661–6679.

Takeda, Y., Sarai, A., and Rivera, V. M. (1989). Analysis of the sequence-specific interactions between Cro repressor and operator DNA by systematic base substitution experiments. *Proc. Natl. Acad. Sci. USA* 86: 439–443.

van Helden, J., André, B., and Collado-Vides, J. (1998). Extracting regulatory sites from the upstream region of yeast genes by computational analysis of oligonucleotide frequencies. *J. Mol. Biol.* 281: 827–842.

Wasserman, W. W., Palumbo, M., Thompson, W., Fickett, J. W., and Lawrence, C. E. (2000). Human-mouse genome comparisons to locate regulatory sites. *Nat. Genetics* 26: 225–228.

Workman, C. T., and Stormo, G. D. (2000). ANN-Spec: A method for discovering transcription factor binding sites with improved specificity. *Pac. Symp. Biocomput.* 5: 464–475.

SUGGESTED READING

Ptashne, M. (1986). *A Genetic Switch*. 2nd ed. Palo Alto, Calif.: Cell Press Blackwell Scientific Publications. Highlights the importance of multiple proteins competing for the same binding sites and the contribution of cooperative binding to regulation.

Stormo, G. D. (1991). Probing the information content of DNA binding sites. *Methods Enzymol.* 208: 458–468. Connects experimental methods for determining specificity with computational models and analyses.

Von Hippel, P. H. (1979). On the molecular bases of the specificity of interaction of transcriptional proteins with genome DNA. In *Biological Regulation and Development*, vol. 1. R. F. Goldberger (ed.). New York: Plenum, pp. 279–347. A great introduction to protein-DNA specificity and the information necessary in regulatory systems.

URLs FOR RELEVANT SITES

AlignACE. `http://atlas.med.harvard.edu/cgi-bin/alignace.pl`

CONSENSUS. `http://ural.wustl.edu/consensus`

Gibbs's sampler. `http://bayesweb.wadsworth.org/gibbs/gibbs.html`

MEME. `http://meme.sdsc.edu/meme/website`

5 Genomics of Gene Regulation: The View from *Escherichia coli*

Julio Collado-Vides, Gabriel Moreno-Hagelsieb, Ernesto Pérez-Rueda, Heladia Salgado, Araceli M. Huerta, Rosa María Gutiérrez, David A. Rosenblueth, Andrés Christen, Esperanza Benítez-Bellón, Arturo Medrano-Soto, Socorro Gama-Castro, Alberto Santos-Zavaleta, César Bonavides-Martínez, Edgar Díaz-Peredo, Fabiola Sánchez-Solano, and Dulce María Millán

The first genome sequence from a free-living organism was completed in 1995. We do not know yet which will be the first genome to be completely deciphered, but *Escherichia coli* is a natural candidate for such a feat. Certainly *E. coli* has been the most studied cell since the early days of molecular biology. Fred Neidhardt, a well-known microbiologist, said once that every biologist knows about at least two cells: the one he is studying and working with, and—even if he is not aware of it—*E. coli*. Nonetheless, there is a large amount of information still missing about the genome of *E. coli*. For instance, currently there is functional information for about half of its 4290 genes. While we know the transcription unit (TU) organization of a third of such genes (around 1400 genes), only about 600 promoters have been experimentally mapped.

Despite being incomplete, the information gathered on *E. coli*, in a sense, offers a benchmark for computational biology. Consider the annotation process of a new genome. This process consists in associating the known and predicted biological properties of segments of DNA, and their products, in the case of genes, with the corresponding region in the genomic sequence. A first step is guessing genes by methods that take into account the statistics of small oligonucleotide distributions based on an initial set of genes. This "seed" is either a group of experimentally known genes or a set of large ORFs (open reading frame, a continuous region of DNA coding for a protein or a fraction of it) found in the genome that by some evidence are likely to be genes.

Typically the seed is obtained by the use of a computer program designed to translate the whole DNA sequence into all possible proteins, then compare them against the database or repository of experimentally known or predicted protein products (see chapter 2). It is here that *E. coli* contributes a large number of experimentally characterized genes and gene products. This chapter shows that in bioinformatics the knowledge accumulated on *E. coli* extends to other goals in the *in silico* description of the biology of gene regulation.

As a result of an initial review of transcriptional regulation in *E. coli* (Collado-Vides et al., 1991), our laboratory has for years been gathering data from the literature and organizing it into a database, RegulonDB (for a detailed description of this database, see Salgado et al., 2001). A relational structure supports its availability on the Web (see `http:/www.cifn.unam.mx/Computational_Genomics/regulondb/`). RegulonDB contains information on transcription initiation as well as operon organization in *E. coli* K-12. Thus, the information it contains can be mapped with its precise coordinate in the chromosome. Table 5.1 summarizes the current contents of the database.

Using the information gathered in RegulonDB, we have implemented several methods to enlarge the known cases and complement them with computational predictions. In this way a complete description of genes in TUs and their associated regulatory elements can be proposed. In this chapter we illustrate how this information can be used in the analysis of global gene expression profiles in *E. coli* grown in different conditions. This implies generating predictions of steady states of a given regulatory network structure. Furthermore, some methods are illustrated that use the experimentally determined transcriptome profiles to generate predictions of the structure of the network. Certainly, we do not yet know all of the interacting molecules defining the network and its dynamics. Finally, we discuss how these methods can be extended to analyze other microbial genomes.

In the first part of this chapter we describe the methods used to predict the structure of the regulatory network using sequence information. We also discuss the use of Bayesian methods to enrich some of these predictions. Next we give some examples of the use of the

Table 5.1 Contents of RegulonDB: Number of Elements and Increase from Previous Years

Object	1997	1998	1999	2000[a]	2001[b]
Regulons	99	83	83	165	166
Regulatory Interactions	533	433	433	642	935
Sites			406	469	750
Products:			4405	4405	4405
Regulatory polypeptides			83	165	318[c]
RNAs			115	115	115
Other polypeptides			4207	3976	3972
Protein complexes	99	83	83	165	166[d]
Genes	542	456	4405	4405	4405
Transcription units	292	230	374	528	657
Promoters	300	239	432	624	746
Effectors	35	36	36	36	66
External References	2050	2011	4394	4704	4943
Synonyms		681	3525	3525	3525
Terminators			40	86	106
RBSs			59	98	133
Conformations				83	201
Conditions				10	10

[a] Number of elements as of 1 October 2000 in RegulonDB.
[b] Number of elements as of 7 November 2001 in RegulonDB.
[c] Total of 318 transcriptional DNA-binding regulators, of which 165 have experimental evidence and the rest have been predicted based on their helix-turn-helix DNA binding motif.
[d] The number of protein complexes decreased from 165 to 163 due to the constant verification of the data. It was found that two pairs of regulatory polypeptides, with experimental evidence, actually bind together to create two single protein complexes. See the glossary for the definition of regulon, operon, and transcription unit.

information in the database in the analysis of transcriptome experiments. Finally, we discuss how these methods are expected to change with the increasing number of related genomes sequenced, and discuss some ideas on the evolutionary origin of transcriptional regulation.

METHODS FOR COMPUTING GENOMIC ELEMENTS OF GENE REGULATION

Regulatory Signals: Promoters and Regulatory Binding Sites

Transcriptional regulatory proteins affect the binding of the RNA polymerase (RNAP) to the promoter, that is, the binding sites of RNAP located upstream of genes. By means of various mechanisms, regulatory proteins can alter any step in the transition involving the binding and the subsequent steps leading to the initiation of a stable elongation complex. Once the RNAP binds to the DNA, it undergoes conformational changes from the so-called closed complex to a conformation that separates the DNA strands into an open complex. Transcription of the first two to seven nucleotides initiates, in a process that can be aborted, causing the unbinding of RNAP from the DNA and the liberation of short oligonucleotides. If transcription continues, RNAP forms a stable elongation complex that will proceed to termination of an mRNA transcript containing one or several genes (for a detailed review of this process, see Mooney et al., 1998).

The amount of mRNA produced depends on the regulation of all possible biochemical transitions involved in the complete process (Gralla, 1990). We limit the discussion here to the regulation at the level of transcription initiation, which depends on both intrinsic factors—strength of RNAP binding to the promoter, speed of transcription initiation—and extrinsic factors. The latter basically consist of regulatory proteins that bind to DNA around the promoter region and either facilitate or prevent transcription initiation.

Transcription initiation is highly specific. The RNAP holoenzyme (RNAP core plus a sigma factor) recognizes the precise site of transcription initiation. In the absence of the sigma factor, the RNAP binds to DNA in a much less specific manner. While molecules are capable of such exact recognition, the challenge is to generate mathematical and computational methods that can mimic this high resolution of recognition. We are far from anything similar. Computational methods to predict promoters and other protein-binding sites, such as operator sites, are known to generate a high number of false positives. Although one can solve this problem by limiting the search to a threshold equal to the average score of known sites, or to the average of such scores minus

A	12	42	27	19	00	05	57	49	43		A	16	21	07	103	28	52	40	05	19
C	36	01	40	07	16	18	36	26	09		C	20	21	14	01	13	21	27	00	15
G	28	32	15	14	19	55	00	00	27		G	44	21	06	02	17	11	19	00	45
T	30	31	24	66	71	28	13	31	27		T	26	43	79	00	48	22	20	101	27

C	A	C	T	T	G	A	A	A	Consensus	G	T	T	A	T	A	A	T	G

-35 Matrix   [15 .... 21]   -10 Matrix

Figure 5.1 Weight matrices of the −10 and −35 boxes of *E. coli* sigma 70 promoter. Consensus matrices were generated using the wconsensus program, This −10 consensus pattern agrees with that reported in the literature. TTGACA is the classic consensus on −35; our −35 differs from that in one base pair. We did not use the classic patterns as seeds to make the matrices, as had been done in previous analyses (Hawley and Mac-Clure, 1983; Harley and Reynolds, 1987; Lisser and Margalit, 1993).

one or two standard deviations (Robison et al., 1998), this solution eliminates weaker sites that could potentially be relevant.

Chapter 4 describes in detail the classic methodology of weight matrices that is widely used to identify DNA binding sites. Figure 5.1 shows the weight matrices of the two conserved elements of *E. coli* sigma 70 promoters, the −10 and the −35 boxes. The average scores of the boxes are 3.12 and 2.17 bits, respectively, which add up to 5.28 bits. This means that this signal might be found randomly every $2^{5.28}$, or around once every 60 base pairs (bp). The computational ability to identify similar putative promoter sites depends on the universe to be searched. Thus, Lukashin et al. (1989) claim that they can identify 99% of promoters, but their search is limited to oligonucleotides of the same size as the promoters. This is useless in searching for promoters in a genome. The significant genomic universe to search for promoters can be estimated by looking at the distance distribution of promoters in relation to the beginning of genes.

Figure 5.2 shows that 90% of all sigma 70 promoters are located within 250 bp upstream of the ATG of the beginning of the gene. These 496 promoters are located in 392 regulatory regions; in other words, 81.6% of known TUs have a single promoter, 13.1% have two promoters, and 5.3% have three or more. Thus, the interesting challenge for computational genomics is that of identifying promoters, usually one, in regions of 200 to 300 bp. When identifying a promoter in a region of 250 bp, the efficiency can drop to 20%. For instance, searching

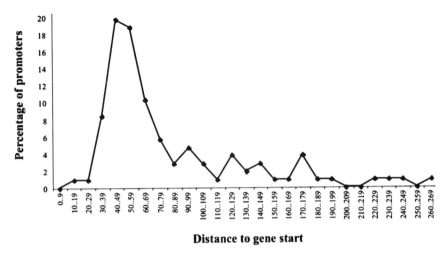

Figure 5.2 Distance distribution of promoters, relative to the beginning of the gene in *E. coli*. Distribution of 496 sigma 70 promoters in relation to the beginning of the downstream gene. Each promoter was positioned based on its −10 box. The highest fraction of promoters falls around 50 bp upstream from the beginning of the gene.

in the whole *E. coli* genome for thresholds equal to the average of scores of experimentally known promoters generates 1341 possible promoters.

If we search using the weight matrices of figure 5.1, using a threshold low enough to make sure we identify 90% of a set of known promoters (commonly the average minus three standard deviations), we get an average of 55 putative promoters per upstream region 250 bp long. That is, there are on average more than 50 additional apparent signals—or noise—mixed with each true signal. In fact, more than 50% of these upstream regions show a putative promoter with a better score than the true known promoter. Figure 5.3a shows the curves of specificity and sensitivity as a function of the threshold. The curve will never rise above 50% sensitivity, since, as just mentioned, more than 50% of regulatory regions show a signal stronger than the true promoter.

No doubt the statistics summarized in a weight matrix are a weak reflection of the molecular ability of RNAP to recognize its binding sites. We have implemented a method that improves the specificity of recognition by an order of magnitude. The method consists of two steps. The first is to obtain a large population of potential candidates, using a low threshold. The second is a "competition" process that

Julio Collado-Vides et al.

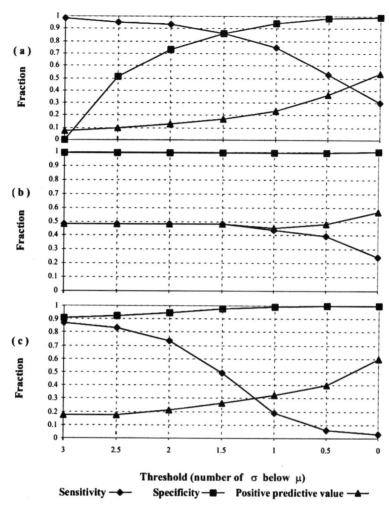

Figure 5.3 Evaluation of detection of promoters by several methods. Promoters were searched in a region 250 bp ustream of the gene. (a) Classic method using consensus/patser. (b) Selecting the strongest scoring site in the region. (c) Using the covering algorithm of competition among putative promoters as described in the text. Sensitivity is defined as the ratio of true positives divided by all true sites (the sum of true positives and false negatives). Specificity is defined as the ratio of true negatives divided by the sum of true negatives plus false positives. The positive predictive value is the ratio of true positives divided by all the sites reported by the method (true positives plus false positives). True sites are those experimentally reported, and positive sites are those found by the method.

selects a few putative promoters based on a comparison of all candidates within the regulatory region. The competition involves the "intrinsic" strength (the added scores of the -10 and -35 boxes, plus the distance separating them) as well as information from the context (i.e., the distance to the beginning of the gene). Interestingly, the method works better when this competition is performed by splitting all signals into groups of candidate promoters with boxes separated by a given distance. All candidate signals are split into seven groups (those whose boxes are separated by a distance of 15 bp, up to those separated by 21 bp), and each group contributes one best candidate.

Although analyses splitting promoters in this way have already been performed (O'Neill and Chiafari, 1989), we initially had no idea of why this independent competition performs better. A plausible explanation is that by means of this mechanism, we are conserving at the end signals that can overlap (since only promoters with different in-between distances can overlap). This would imply that the precise recognition of a promoter by RNAP is achieved by multiple promoter signals located close together, overall providing a higher energy of binding (see chapter 4). Our method yields a true promoter out of six candidates on average (with 80% efficiency; see figure 5.3b), compared with one out of 50 if selected at random.

The computational analyses of operator-binding sites for specific transcriptional regulators around the promoter region follows the same basic strategy of weight matrices that is applied to upstream regions in a genome, but the number of known sites is fewer. RegulonDB currently accounts for around 160 DNA-binding transcriptional regulators with experimental evidence. As of May 2001, we have been able to gather experimental information on known binding sites for 62 proteins. We can construct specific weight matrices (with at least four sites) for 35 regulators and search for potential new sites in upstream regions in the genome. This time, given the distribution of sites located around 200 bp upstream from the promoter initiation site (Gralla and Collado-Vides, 1996), we have to search in a region of around 400 bp upstream from the beginning of the gene. This interval is obtained by adding the distance from the promoter to the beginning of the gene, and the distance from the sites to the promoter.

We did not consider promoter prediction to be reliable enough to limit the search for sites to the vicinity of the predicted promoters. In fact, if one looks at the information content of all regulatory proteins,

Julio Collado-Vides et al.

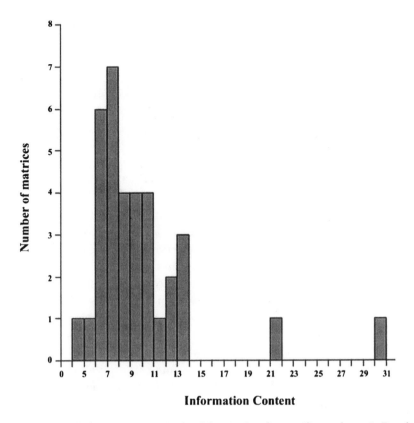

Figure 5.4 Information content of weight matrices for specific regulators in RegulonDB. The information content was obtained using consensus with the reported binding sites of 35 regulatory proteins. The values were adjusted against the sample size. For instance, two of the matrices obtained had less than 6 bits of information, and only one, corresponding to ArcA, has 30 bits of information.

the gradient goes from around 6 to 28 bits (figure 5.4). This indicates that the promoter site (−35 plus −10), with around six bits, is on the lower range of information content, compared with sites for transcriptional regulators. Assuming that this information content reflects the energy of binding, and therefore the equilibrium constant of the association between the protein and the DNA sites, it would make sense to see a lower constant for RNAP, given the greater abundance of RNAP compared with that of specific regulators (Robison et al., 1998). Searching in the set of upstream regions of all genes in the genome, we find more than 44,000 putative CRP sites, whereas for LexA we find only 160 sites (Thieffry, Salgado, et al., 1998). These numbers illustrate the problem of high false positives or poor specificity in the search for operator sites for transcriptional regulators. Furthermore, as mentioned before, we have sites for only around one sixth of all transcriptional regulators. A way to considerably diminish the false positives is by restricting the predicted sites to those located at positions relative to the beginning of transcription similar to positions of known sites. This can be achieved by detailed analysis of the combination of sites into modules or "phrases" (Rosenblueth et al., 1996; Frech et al., 1997).

So far we have focused on methods that deal with sensors that are built from a set of known binding sites for a given protein. The availability of transcriptome data has motivated the implementation of different types of methods aimed at identifying common motifs in sets of regulatory regions of genes observed to be coexpressed (Goffeau, 1998). Although programs in this direction were implemented years ago (Waterman et al., 1984; Stormo, 1990), the recent boom of transcriptome and chip methodologies has motivated new methodologies such as Gibbs sampling (Lawrence et al., 1993) and others (see chapter 6). We have implemented a method that searches exhaustively in all the space of sequences, and identifies "words" or oligonucleotides up to seven bases long that are overrepresented in a given family of upstream regions (van Helden et al., 1998). Its performance has been similar to several other methods when analyzing a set of coexpressed genes in yeast. Later, a variation of this basic idea was implemented that allows for the search for words with internal symmetry, with direct or inverted repeats, or palindromes. This program is called dyad-detector (van Helden et al., 2000).

More recently we have tested the performance of dyad-detector against sets of coregulated genes obtained from RegulonDB. An interesting observation in the first results enabled us to develop a strategy that improves the accuracy of the method in detecting genes regulated by a given protein, and to identify the binding site. This method starts with a family of upstream regions containing the set of binding sites for the known or unknown protein (if the set comes from coexpressed genes, for instance). We then use the dyad-detector to identify the overrepresented words. Most of these significant words accumulate near true binding sites (figure 5.5). Focusing on islands of overlapping words or motifs offers an additional step to increase the specificity of the method.

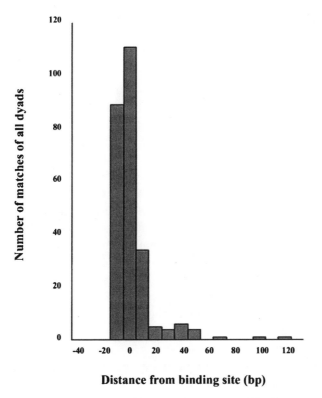

Distance from binding site (bp)

Figure 5.5 Distances from binding site of all dyads detected at the upstream regions of genes regulated by LexA. Dyads with a significance higher than 0 were obtained with the dyad-detector program (van Helden et al., 2000). Their location in relation to the true LexA sites is shown, with negative distances corresponding to those that overlap true sites.

With additional information coming from other sources that define groups of genes or operons as plausibly coregulated sets, we may be able to predict, in principle, the binding sites for a set close to the complete collection of 314 transcriptional regulators in *E. coli* (known and predicted; see next section). This will be an important step forward in the characterization of the structure of the network of transcriptional interactions of the whole cell. We can imagine having 314 sets of binding sites distributed in the upstream regions of the genome, and defining the corresponding weight matrices that expand the currently known 50 or so weight matrices.

We turn now to the analysis of transcriptional activators and repressors that affect transcription initiation. Helix-turn-helix (HTH) is the dominant motif used by these specific regulators to bind to DNA in bacteria, although other motifs, such as beta-sheet and zinc fingers, are found in a few cases. According to our estimates, more than 90% of bacterial regulators involve an HTH motif (Pérez-Rueda and Collado-Vides, 2000). Similar methods, based on weight matrices (profiles) and regular expressions (pattern search, i.e., prosite patterns), can be used to predict transcriptional regulators identifying the conserved HTH motif by means of sequence similarity. In this way the set of 160 known regulators expands to a total of 314 transcriptional regulators in *E. coli*. These regulators are grouped in 24 different families, which are assumed to be groups of proteins with a common evolutionary origin. The functional roles of these families can be inferred from the observation that repressors have their HTH motif located predominantly at the N-terminus, and activators have it at the C-terminus, of the protein (Pérez-Rueda et al., 1998; see figure 5.6). This is another useful piece of information that contributes to the solution of the problem of predicting the whole network of regulatory interactions involved in transcription initiation.

Assume that we have identified an equal number of families of protein-binding motifs as we have of transcriptional regulators. The problem then is to identify which protein binds to which set of sites, and thus to identify which protein regulates which set of genes. Ideally, we would like to directly identify the sequence of the operator site from the sequence of the DNA-binding motif of the regulator, as has been done for the GalS protein (Müller-Hill and Kolkhof, 1994). However, with the data in RegulonDB we have verified that this striking observation is indeed an exceptional case. A genomic approach putting together several pieces of information seems reasonable. Some of these pieces and their contribution to limiting the association of the regulator and its corresponding regulatory sites and regulated genes are the following:

· Regulators are predicted as activators or repressors, and are also predicted as members of a particular family of evolutionarily related proteins. We have observed that families tend to regulate genes of relatively similar function (Pérez-Rueda and Collado-Vides, 2000).

· The position of the binding site in relation to the regulated promoter can corroborate if it is an activator or a repressor. Even in the case of

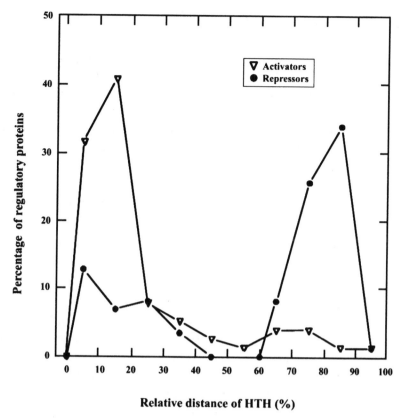

Figure 5.6 Relative location of helix-turn-helix (HTH) motifs within regulatory proteins. The total length in amino acids of each regulatory protein was normalized to 100, and the relative position of the center of the HTH motif for each protein was obtained (see Pérez-Rueda et al., 1998).

dual proteins, we have observed that for a given specific regulator, when it works as an activator, its sites are closer to the promoter, compared with the positions of the sites used to repress (Gralla and Collado-Vides, 1996).

• Transcriptome snapshots of a regulated transition would in principle enable the search for correlations in the expression profiles of regulatory and regulated genes, and in this way contribute to restricting their identification.

• Sets of genes subject to common regulation, or regulons, can be predicted on the basis of phylogenetic profiles (Pellegrini et al., 1999; see also chapter 6), and on the basis of expanding the method of operon prediction described below.

• We know that regulatory proteins are autoregulated in a good number of cases (Thieffry, Huerta, et al., 1998). Thus, finding a motif present upstream of a regulatory protein gene would suggest that all genes having similar sites are targets of the regulator.

A clustering approach can be implemented that uses all this information and that is capable of grouping genes on the basis of both knowledge from the database and data from transcriptome experiments. We have implemented, and are currently evaluating, a clustering method based on Bayesian statistics that is able to perform such tasks.

Operons

In bacteria, genes are clustered in monocistronic and polycistronic (operons) TUs. The information available in RegulonDB for more than 400 TUs with experimental evidence enabled us to implement a method that can predict operons. This method is based on the clearly different distribution of intergenic distances of pairs of genes in the same operon, as opposed to pairs at the boundaries of TUs. It also uses the conservation of functional class of genes within an operon to predict operons in the rest of the genome (Salgado et al., 2000).

The organization of all genes into TUs has motivated the definition of the set of "minimal upstream regions" (MURs) of the genome as the set of upstream regions of all TUs in the genome. This clustering simplifies the problems of motif searching and of the architecture of the network of interactions. Forty-two hundred genes and ORFs cluster in around 2600 TUs, and thus there are 2600 MURs.

We have shown that the set of TUs in a given regulon can be expanded by the use of comparative genomics combined with evidence of upstream regulatory motifs (Tan et al., 2001). In this example we show that "orthologue TUs," found in *E. coli* and *Haemophilus influenzae*, can be added to a regulon if such TUs have a similar binding site. An expansion of this idea, as suggested by Galperin and Koonin (2000), takes advantage of the high frequency of rearrangement of operons. It is expected that such rearrangements of the gene elements of operons occur among related TUs (regulons). Starting with a set of genes known to be part of a regulon in *E. coli*, one would need to find the orthologues

Julio Collado-Vides et al.

in other genomes. If such orthologues are associated in an operon with another gene, and that gene is also present in *E. coli*, there would be a chance that this gene is part of the same regulon. Some evidence that the rearrangements occur among related operons has been published (Lathe et al., 2000).

A complete analysis would need TU predictions, in different genomes, to start with. We have expanded the method originally developed in *E. coli* and have found evidence suggesting that it works well in any prokaryotic genome. It will be very interesting to extrapolate some of these pieces of the puzzle of genomic regulation and apply them to different bacterial genomes. Certainly the role of *E. coli* as an oasis of annotations, compared with much less studied related microorganisms, must be extrapolated further (see more about these predictions in the section "Comparative Genomics and Evolution of Gene Regulation").

TRANSCRIPTOME ANALYSES

All the accumulated experimental knowledge and computational predictions of gene clustering in operons and gene regulation provide valuable information for analyzing expression profiles from transcriptome experiments in *E. coli*. The analysis of any given condition requires the identification of the function of the set of affected—induced or repressed—genes, the organization of such genes into operons, their regulatory regions with promoters and regulatory sites; that is, the gathering of all known information on such genes. There is also a need to have at hand methods that can organize and manipulate the information. We have implemented interfaces that integrate outputs of programs that search in RegulonDB and display integrated information about the regulation of gene expression of a given set of genes (see `http://www.cifn.unam.mx/Computational_Genomics/GETools`).

Several interesting questions can be raised when comparing the data that have accumulated through years of experimentation with data obtained in a single transcriptome. One question is to evaluate and understand how congruent the observed profile is to what we know about the network of interactions. At a first, broad level of analysis, statistical evaluations of such congruence can be obtained. Genes grouped in operons tend to be subject to a common mechanism of regulation. Thus, most of the time one would expect that under any growth con-

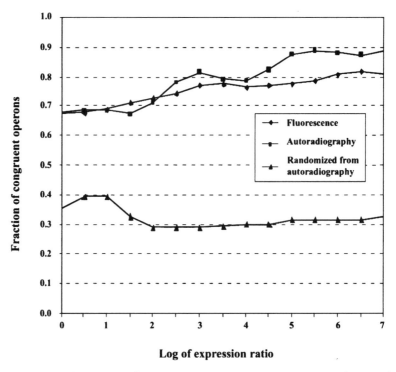

Figure 5.7 Congruence of operons in transcriptome experiments. Evaluation of congruence of transcriptome experiments as measured by the number of operons completely expressed or repressed by the number of operons represented at different levels of induction or repression (here shown as absolute value). The example shows the heat shock experiments measured by fluorescence and by autoradiography.

dition, all genes within an operon will be either induced or repressed. We define such operons as congruent. We could count the fraction of congruent operons as a function of the threshold of gene expression.

If the low expression values get mixed with the noise of detection, one would expect a curve of increasing congruence as the threshold rises. This is what we observe in several transcriptome experiments, one of which is illustrated in figure 5.7. Although such a congruent behavior is not surprising, it is a useful index, based on the biology of *E. coli* genes and operons, to use as an "internal control" of experiments across different conditions. More interestingly, if the high region of congruence of this curve has some biological meaning, it suggests that for a large number of changes in growth conditions, *E. coli* adapts through the regulation of an important number of genes, no fewer than 100 or 150. This clearly suggests mechanisms for concerted regulation of a number of genes that is well beyond the average number of genes in a single known regulon.

There is enough information to analyze the congruence of cell behavior in more detail and from a different perspective. Snapshots of the expression flexibility of the cell are like photographs taken from a boat at high speed—the picture can easily be blurred. The set of regulatory interactions in a cell forms a network in which the level of expression of one gene can easily affect that of others, either directly or indirectly. In order to evaluate the congruence of genes within such networks, we need to consider all known interactions. This type of analysis, at a given moment in time, can be called "internal congruence" or "synchronous congruence"—as opposed to the analysis of a transition of the cell when changing the conditions of growth.

We investigated whether groups of genes known to be precisely coregulated (by one or several proteins, but exactly the same set) show homogeneous patterns of expression in transcriptome data. We concentrated on the study of absolute levels of expression of *E. coli* grown at 37 °C in minimal medium, since we have data from the experiment performed four times. These experiments were performed in the laboratory of Fred Blattner at the University of Wisconsin, Madison. Defining expression levels as either *on* or *off*, based on their absolute values, we can identify subsets of coregulated genes that behave in a homogeneous way—either *on* or *off*. A binomial analysis with a priori probabilities p and $(1 - p)$ for *on* and *off*, based on their frequency in the complete set of genes in the experiments, shows the statistical significance of these results. A set of genes coregulated by a single protein and with a statistically significant behavior is shown in table 5.2. This may not be the case for all coregulated groups. More elaborated analyses will surely be required to deal with the flexible response of regulons.

In order to perform similar analyses, this time comparing transcriptome experiments on *E. coli* in different conditions, we need to employ a computational machinery that will simulate direct and indirect effects and their state transitions. In a way, the ideas presented here are rather simple in terms of simulation of networks. The recent literature addressing the analysis of transcriptome data and modeling of regulatory networks is growing very fast. (See, for instance, the

Table 5.2 Internal congruence in gene expression of coregulated genes

Co-regulating proteins			Data						
			Experiments						
			Gene	1	2	3	4	Off	On
CsgD	CsgG	OmpR	csgD	*off*	*off*	*off*	*off*	4	0
CsgD	CsgG	OmpR	csgE	*off*	*off*	*off*	*off*	4	0
CsgD	CsgG	OmpR	csgF	*off*	*off*	*off*	*off*	4	0
CsgD	CsgG	OmpR	csgG	*off*	*off*	*off*	*off*	4	0
Totals *on*				0	0	0	0		
Totals *off*				4	4	4	4	16	0
Summary									
% genes *off*			1.00						
% genes *on*			0.00						
Probability			0.01						

An example of a group of genes known to be precisely coregulated by three regulatory proteins: CsgD, CsgG, OmpR. The group shows a homogeneous pattern of expression in repetitions (experiments 1, 2, 3, and 4) of transcriptome data. The expression levels are described as either *on* or *off* pattern, for which we analyzed the statistical significance using a binomial distribution model.

approaches described in chapters 7, 8, and 11, as well as references and suggested literature cited there.)

COMPARATIVE GENOMICS AND EVOLUTION OF GENE REGULATION

Since the work by Zuckerkandl and Pauling (1963), we have been aware of the historical content within macromolecular sequences. Conservation in the course of evolution is becoming a powerful source of information for a new generation of computational methods that use full genomic sequence and information. For instance, predicting regulatory elements in eukaryotic genomes was recognized until very recently as a difficult computational problem. Although their position in relation to the beginning of the genes is much more flexible than in prokaryotes and allows remote positions, one would expect, when comparing two related organisms, that their functional sites are visible as small, conserved regions as opposed to the rest of noncoding DNA.

This identification is what is called a "phylogenetic footprint" (Fickett and Wasserman, 2000; McCue et al., 2001) in an analogy to chemical footprinting of operator sites.

We have mentioned that regulatory proteins can be grouped into families with a common evolutionary origin. The interesting correlation between the position of the HTH motif and the role in regulation of the protein can be either the result of a convergent process in evolution due to physical constraints, or a trace of the common origin of these proteins. To address this issue, we expanded our data set of transcriptional regulators, and compared them with all ORFs in sequenced bacterial and archaeal sequenced genomes. Our accumulated evidence points to a common origin of this positional restriction (Pérez-Rueda and Collado-Vides, 2001). Certainly, when the HTH conserved sequence within a family is analyzed, it expands beyond the strict functional, characterized HTH motif, suggesting that a trace of similar sequence with no binding task is still visible. Second, not only evolutionary families tend to have conserved positions; the distribution of families across the different microbial genomes is such that we can estimate the point in evolution when they emerged. A quite interesting observation in this analysis is a set of four repressor families that show homologues in archaea, indicating that this motif existed before the divergence, around three billion years ago (Feng et al., 1997), between the current bacteria and the archaea.

The evolutionary trace in operon organization is also rather striking in the microbial world. The organization of operons in E. coli has shown, as already mentioned, genes within operons with a strong preference for short intergenic distances as opposed to a flat, almost equiprobable distribution for a range of distances between genes at the boundaries of TUs (figure 5.8). This is a strict genomic observation in the sense of a property that can be analyzed only when complete genomes are available. The intergenic distances aid in predicting functional relationships between genes—because genes in the same operon tend to participate in the same function—without any need for homology detection among proteins.

In order to evaluate whether the same method can be applied in predicting operons in bacteria other than E. coli, one needs known sets of experimentally characterized operons in other bacteria. Fortunately, a collection of experimentally characterized operons is available for Bacillus subtilis, a gram-positive bacterium. To our surprise,

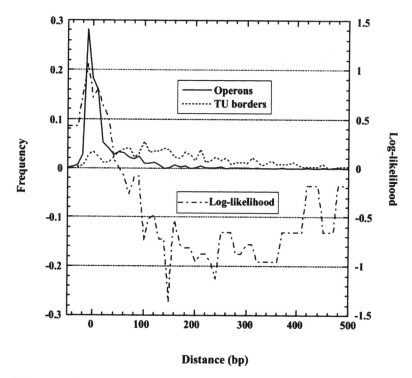

Figure 5.8 Intergenic distances of genes within operons and at TU borders. Distribution of intergenic distances of genes within operons (continuous line), and of genes at the borders of transcription units (discontinuous line, upper part). The figure also displays the log-likelihoods derived from the comparison of the frequency of pairs of genes within operons separated by a given distance versus that among genes at TU borders (see Salgado et al., 2000). Intergenic distances were obtained from known operons, and distances at borders were obtained from transcription units (i.e., operons as well as genes transcribed in isolation).

the predictive accuracy of our method for this set is as high as that for the operons in RegulonDB, using the parameters defined with the *E. coli* collection.

Still, evidence is necessary to show that the method is applicable to any prokaryotic genome. Another observation about the properties of genes within experimentally known operons, compared against genes at TU boundaries, is that they have a higher tendency to be conserved as neighbors in other genomes (figure 5.9). We tested to see if this holds true for predicted genes within operons and predicted boundaries (Moreno-Hagelsieb et al., 2001). This independent test gave the ex-

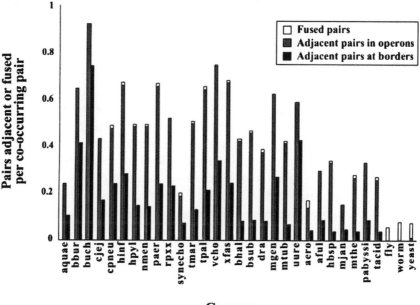

Figure 5.9 Conservation of *E. coli* pairs of genes inside operons and at their border. The first column in each organism represents pairs in operons, while the second represents genes at the borders of transcription units. The column representing pairs in operons is always higher, and fusions occur only among conserved genes corresponding to operons. The labels mostly correspond to the file names at GENBANK: aquae, *Aquifex aeolicus*; bbur, *Borrelia burgdorferi*; buch, *Buchnera* sp. APS; cjej, *Campylobacter jejuni*; cpneu, *Chlamydia pneumoniae*; hinf, *Haemophilus influenzae* Rd; hpyl: *Helicobacter pylori* 26695; nmen, *Neisseria meningitidis* MC58, paer, *Pseudomonas aeruginosa*; rpxx, *Rickettsia prowazekii*, synecho, *Synechocystis* PCC6803; tmar, *Thermotoga maritima*; tpal, *Treponema pallidum*; vcho, *Vibrio cholerae*; xfas, *Xylella fastidiosa*; bhal, *Bacillus halodurans*; bsub, *Bacillus subtilis*; dra, *Deinococcus radiodurans*; mgen, *Mycoplasma genitalium*; mtub, *Mycobacterium tuberculosis*; uure, *Ureaplasma urealyticum*; aero, *Aeropyrum pernix*; aful, *Archaeoglobus fulgidus*; hbsp, *Halobacterium* sp.; mjan, *Methanococcus jannaschii*; mthe, *Methanobacterium thermoautotrophicum*; pabyssi, *Pyrococcus abyssi*; tacid, *Thermoplasma acidophilum*; fly, *Drosophila melanogaster*; worm, *Caenorhabditis elegans*; yeast, *Saccharomyces cerevisiae* (see Moreno-Hagelsieb et al., 2001).

pected results, that is, the method derives predicted genes within operons that are conserved as neighbors more frequently than predicted boundaries. Other tests confirm the applicability of the method, such as the conservation of the peak observed in the distance distributions of all genes transcribed in the same direction. The peak is always located at the same place, and it is a reflection of the abundance of operons of each genome. This information on abundance can be used to adjust the predictions for greatest accuracy, but in principle the method performs surprisingly well in any prokaryotic genome.

In summary, we started with an expected observation: that pairs of genes in the same operon tend to be close to each other, in contrast to pairs of genes at the boundaries between two neighboring TUs. The difference in the distribution is so clear that this single parameter suffices to predict operons with an accuracy higher than 80% in *E. coli*. A similar accuracy in the set of known operons in *Bacillus* is rather surprising, given the evolutionary distance between these organisms. Also important is the different evolutionary conservation of pairs of genes in operons, as opposed to genes at the boundaries of TUs. The resulting methodology enables a quite accurate prediction of TUs in all microbial genomes. This computational ability is based on a universal conservation of the basic architecture of operons within the microbial world.

CONCLUSIONS

Many questions are open, and some of them may remain mysteries for a long time. Transcription regulation as a network of interactions clearly illustrates the complexity of biological architectures. We do not know how deeply we will be able to understand such networks with current scientific approaches. There is no clear quantitative method to address the emergent properties of complex biological systems. These limitations bring a healthy skepticism to what we will be able to do (Krischner et al., 2000). However, the defined phenotypes of individual mutations in complex differentiation pathways, such as aging, suggests that complex as biology may be, it is decomposable and subject to analysis.

Transcriptional regulation and operon organization in bacteria bring us to a different point in the history of the biological complexity, the one of the origin of life, or at least the origin of cellular life and its early speciation in what became the archaeal, bacterial, and eukaryotic domains of life. *E. coli* or another bacterium may be the first completely annotated genome in the near future.

It is true that in several aspects bacteria are simpler systems, compared to higher organisms. But this simplicity does not prevent novel methods and representations that can integrate the biology of a bacterial cell from being equally useful when studying eukaryotic cells. One further step will be to study gene regulation and the associated metabolic capabilities of the regulated genes in a single integrated model.

ACKNOWLEDGMENTS

We acknowledge Víctor del Moral's computational support in the laboratory. The research summarized here was supported by grant 0028 from Conacyt.

REFERENCES

Blattner, F. R., Plunkett, G. III, Bloch, C. A., Perna, N. T., Burland, V., Riley, M., Collado-Vides, J., Glasner, J. D., Rode, C. K., Mayhew, G., Gregor, J., Davis, N. W., Kirkpatrick, H. A., Goeden, M. A., Rose, D. J., Mau, B., and Shao, Y. (1997). The complete genome sequence of *Escherichia coli* K-12. *Science* 277: 1453–1462.

Collado-Vides, J., Magasanik, B., and Gralla, J. D. (1991). Control site location and transcriptional regulation in *Escherichia coli*. *Microbiol. Rev.* 55: 371–394.

Feng, D. F., Cho, G., and Doolittle, R. F. (1997). Determining divergence times with a protein clock: Update and reevaluation. *Proc. Natl. Acad. Sci. USA* 94: 13028–13033.

Fickett, J. W., and Wasserman, W. W. (2000). Discovery and modeling of transcriptional regulatory regions. *Curr. Opinion Biotechnol.* 11: 19–24.

Frech, K., Danescu-Mayer, J., and Werner, T. (1997). A novel method to develop highly specific models for regulatory units detects a new LTR in GenBank which contains a functional promoter. *J. Mol. Biol.* 270: 674–687.

Galperin, M. Y., and Koonin, E. V. (2000). Who's your neighbor? New computational approaches for functional genomics. *Nat. Biotechnol.* 18(6): 609–613.

Goffeau, A. (1998). Genomic-scale analysis goes upstream? *Nat. Biotechnol.* 16: 907–908.

Gralla, J. D. (1990). Promoter recognition and mRNA initiation by *Escherichia coli* E sigma 70. *Methods Enzymol.* 185: 37–54.

Gralla, J. D., and Collado-Vides, J. (1996). Organization and function of transcription regulatory elements. In F. C. Neidhardt, R. Curtiss III, J. Ingraham, E. C. C. Lin, K. B. Low, B. Magasanik, W. Reznikoff, M. Schaechter, H. E. Umbarger, and M. Riley (eds.), *Cellular and Molecular Biology: Escherichia coli and Salmonella*. 2nd ed. Washington, D.C.: American Society for Microbiology, pp. 1232–1245.

Harley, C. B., and Reynolds, R. P. (1987). Analysis of *E. coli* promoter sequences. *Nucleic Acids Res.* 15(5): 2343–2361.

Hawley, D. K., and McClure, W. R. (1983). Compilation and analysis of *Escherichia coli* promoter DNA sequences. *Nucleic Acids Res.* 11(8): 2237–2255.

Hertz, G. Z., Hartzell, G. W. III, and Stormo, G. D. (1990). Identification of consensus patterns in unaligned DNA sequences known to be functionally related. *Comput. Appl. Biosci.* 6: 81–92.

Hertz, G. Z., and Stormo, G. D. (1999). Identifying DNA and protein patterns with statistically significant alignments of multiple sequences. *Bioinformatics* 15: 563–577.

Krischner, M., Gerhart, J., and Mitchison, T. (2000). Molecular "vitalism." *Cell* 100: 79–88.

Lathe, W. C., Snel, B., and Bork, P. (2000). Gene context conservation of a higher order than operons. *Trends Biochem. Sci.* 25(10): 474–479.

Lawrence, C. E., Altschul, S. F., Boguski, M. S., Liu, J. S., Neuwald, A. F., and Wootton, J. C. (1993). Detecting subtle sequence signals: A Gibbs sampling strategy for multiple alignment. *Science* 262: 208–214.

Lisser, S., and Margalit, H. (1993). Compilation of *E. coli* mRNA promoter sequences. *Nucleic Acids Res.* 21(7): 1507–1516.

Lukashin, A. V., Anshelevich, V. V., Amirikyan, B. R., Graverov, A. I., and Frank-Kamenetskii, M. D. (1989). Neural network models for promoter recognition. *J. Biomol. Struct. Dynamics* 6: 1123–1133.

McCue, L. A., Thompson, W., Carmack, C. S., Ryan, M. P., Liu, J. S., Derbyshire, V., and Lawrence, C. E. (2001). Phylogenetic footprinting of transcription factor binding sites in proteobacterial genomes. *Nucleic Acids Res.* 29: 774–782.

Mooney, R. A., Artsimovitch, I., and Landick, R. (1998). Information processing by RNA polymerase: Recognition of regulatory signals during RNA chain elongation. *J. Bacteriol.* 180: 3265–3275.

Moreno-Hagelsieb, G., Treviño, V., Pérez-Rueda, E., Smith, T. E., and Collado-Vides, J. (2001). Transcription unit conservation in the three domains of life: A perspective from *Escherichia coli*. *Trends in Genetics* 17: 175–177.

Müller-Hill, B., and Kolkhof, P. (1994). DNA recognition and the code. *Nature* 369: 614.

Neidhardt, F. C., et al. (eds.). (1987). *Escherichia coli and Salmonella typhimurium. Cellular and Molecular Biology*. Washington, D.C.: American Society for Microbiology.

Neidhardt, F. C., et al. (eds.). (1996). *Escherichia coli and Salmonella typhimurium. Cellular and Molecular Biology*. 2nd ed. Washington, D.C.: American Society for Microbiology.

O'Neill, M. C., and Chiafari F. (1989). Escherichia coli promoters. II. A spacing class-dependent promoter search protocol. *J. Biol. Chem.* 264: 5531–5534.

Pellegrini, M., Marcotte, E. M., Thompson, M. J., Eisenberg, D., and Yeates, T. O. (1999). Assigning protein functions by comparative genome analysis: Protein phylogenetic profiles. *Proc. Natl. Acad. Sci. USA* 96: 4285–4288.

Pérez-Rueda, E., and Collado-Vides, J. (2000). The repertoire of DNA-binding transcriptional regulators in *Escherichia coli*. *Nucleic Acids Res.* 28: 1838–1847.

Pérez-Rueda, E., and Collado-Vides, J. (2001). Common history at the origin of the position-function correlation in transcriptional regulators in archaea and bacteria. *J. Mol. Evol.* 53: 172–179.

Pérez-Rueda, E., Gralla, J. D., and Collado-Vides, J. (1998). Genome analysis and the transcription machinery. *J. Mol. Biol.* 275(2): 165–170.

Robison, K., McGuire, A. M., and Church, G. M. (1998). A comprehensive library of DNA-binding site matrices for 55 proteins applied to the complete *Escherichia coli* K-12 genome. *J. Mol. Biol.* 284: 241–254.

Rosenblueth, D. A., Thieffry, D., Huerta, A. M., Salgado, H., and Collado-Vides, J. (1996). Syntactic recognition of regulatory regions in *Escherichia coli*. *Comput. Appl. Biosci.* 12: 415–422.

Salgado, H., Moreno-Hagelsieb, G., Smith, T. F., and Collado-Vides, J. (2000). Operons in *Escherichia coli*: Genomic analyses and predictions. *Proc. Natl. Acad. Sci. USA* 97: 6652–6657.

Salgado, H., Santos-Zavaleta, A., Gama-Castro, S., Millán-Zárate, D., Díaz-Peredo, E., Sánchez-Solano, F., Pérez-Rueda, E., Bonavides-Martínez, C., and Collado-Vides, J. (2001). RegulonDB (version 3.2): Transcriptional regulation and operon organization in *Escherichia coli* K-12. *Nucleic Acids Res.* 29: 72–74.

Stormo, G. (1990). Consensus patterns in DNA. *Methods Enzymol.* 183: 211–221.

Tan, K., Moreno-Hagelsieb, G., Collado-Vides, J., and Stormo, G. D. (2001). A comparative genomics approach to prediction of new members of regulons. *Genome Res.* 11: 566–584.

Thieffry, D., Huerta, A. M., Pérez-Rueda, E., and Collado-Vides, J. (1998). From specific gene regulation to genomic networks: A global analysis of transcriptional regulation in *Escherichia coli*. *BioEssays* 20: 433–440.

Thieffry, D., Salgado, H., Huerta, A. M., and Collado-Vides, J. (1998). Prediction of transcription regulatory sites in the complete genome of *Escherichia coli*. *Bioinformatics* 14: 391–400.

van Helden, J., André, B., and Collado-Vides, J. (1998). Extracting regulatory sites from the upstream region of yeast genes by computational analysis of oligonucleotide frequencies. *J. Mol. Biol.* 281(5): 827–842.

van Helden, J., Ríos, A. F., and Collado-Vides, J. (2000). Discovering regulatory elements in non-coding sequences by analysis of spaced dyads. *Nucleic Acids Res.* 28(8): 1808–1818.

Waterman, M., Arratia, R., and Galas, D. J. (1984). Pattern recognition in several sequences: Consensus and alignment. *Bull. Math. Biol.* 46: 515–527.

Zuckerkandl, E., and Pauling, L. (1963). Molecules as documents of evolutionary history. *J. Theor. Biol.* 8: 357–366.

SUGGESTED READING

Coe, M. (1992). *Breaking the Maya Code*. New York: Thames and Hudson. A summary of lessons from anthropologists drawn from deciphering ancient human languages. Deciphering genomes is not that different, is it?

Collado-Vides, J., Magasanik, B., and Smith, T. F. (eds.). (1996). *Integrative Approaches to Molecular Biology:* Cambridge, Mass.: MIT Press. A book that expands from databases to regulation, metabolism, and evolution.

Jacob, F. (1997). *La Souris, la Mouche, et l' Homme*. Paris: Editions Odile Jacob. We do not know if there is an English version of this update of *La Logique du Vivant* by the same author, summarizing his view of the recent history of molecular biology. It is worth reading.

URLs FOR RELEVANT SITES

E. coli genome project in Madison, Wisconsin. `http://www.genome.wisc.edu`

RegulonDB access site and also access site laboratory of Program of Computational Genomics, UNAM.
`http://www.cifn.unam.mx/Computational_Genomics/regulondb`

Julio Collado-Vides et al.

6 Discovery of DNA Regulatory Motifs

Abigail Manson McGuire and George M. Church

Short, conserved sequence elements, or DNA motifs, located upstream of the transcriptional start site are often binding sites for transcription factors that regulate a group of genes involved in similar cellular functions. Additional genes with this motif in their upstream regions are good candidates for involvement in the same cellular function. Thus, upstream regulatory motifs can provide powerful hypotheses about links in the genetic regulatory network. These upstream regulatory motifs can be discovered computationally by local alignment of upstream regions from coregulated sets of genes. Here we will focus on the AlignACE motif-finding algorithm (Roth et al., 1998).

Figure 6.1 illustrates the process of motif discovery. In order to discover a motif computationally by using local alignment methods, an accurate list of coregulated genes, or a regulon, is needed. The AlignACE program can tolerate some noise in the alignment process from extra sequence that does not contain the motif of interest, but too much extra sequence will prevent the motif from being aligned. Thus, we must employ additional experimental and computational methods to obtain accurate regulon predictions.

Figure 6.2 illustrates several methods that can be used for regulon prediction. Each of these methods will be discussed in more detail below. mRNA expression data have proved to be very useful for predicting coregulated sets of genes. Genes that are upregulated or downregulated in a single experimental condition are likely to be coregulated, as are genes with similar mRNA expression profiles across many conditions or time points. There are also several computational methods for predicting regulons that are based on comparative genomics. Here we will discuss several methods for predicting regulons, motif discovery algorithms, parameters for evaluating the

Predict coregulated
sets of genes

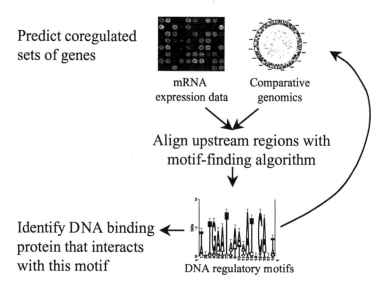

mRNA
expression data

Comparative
genomics

Align upstream regions with
motif-finding algorithm

Identify DNA binding ←
protein that interacts
with this motif

DNA regulatory motifs

Figure 6.1 Process of motif discovery. The first step is to predict coregulated groups of genes, or regulons. This can be done either using expression data from microarrays or by computational methods based on comparative genomics. Next, the upstream regions of genes within the predicted regulons are aligned, using a motif discovery algorithm. The resulting DNA regulatory motifs can be illustrated using a motif logo, as shown here (Schneider and Stevens, 1990). The height of a stack of letters is proportional to the information content, and the relative frequency of each base is given by its relative height. The presence of a significant regulatory motif found in this manner can be used to refine the contents of the predicted regulons, because the presence of a regulatory motif provides additional evidence that this set of genes is coregulated. Experiments can also be designed to test the function of the regulatory motif and to identify the transacting factor that binds to it.

significance of a proposed regulatory motif, and applications of motif finding using AlignACE in bacterial genomes and in *Saccharomyces cerevisiae*. An additional Web tutorial on topics covered in this chapter is available at `http://arep.med.harvard.edu/labgc/amcguire/IBSS/index.html`.

REGULON PREDICTION

Groups Derived from mRNA Expression Data

A powerful method for predicting regulons is to look for patterns in mRNA expression data from DNA chips or microarrays. This has been

A. M. McGuire and G. M. Church

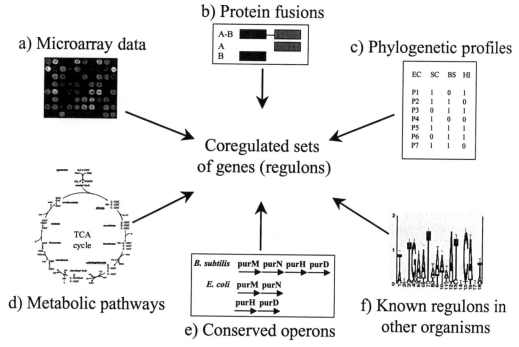

Figure 6.2 Methods for regulon prediction. Method (a) relies on experimental mRNA expression data from microarrays. Methods (b), (c), and (e) are comparative genomics methods based purely on sequence information. Methods (d) and (f) incorporate known biology of well-studied organisms.

done for several sets of experimental conditions in *S. cerevisiae* (Tavazoie et al., 1999; Roth et al., 1998). One particularly effective method for obtaining coregulated sets of genes from mRNA expression data is to cluster genes on the basis of their expression profiles over a large number of experimental conditions, or over a large number of time points (Tavazoie et al., 1999; Wen et al., 1998; Eisen et al., 1998; Tamayo et al., 1999). Different algorithms for clustering such data are reviewed in Sherlock (2000). Many known motifs, as well as new motifs, have been found in the upstream regions of *S. cerevisiae* genes clustered by this method (Tavazoie et al., 1999). A database of expression data in *S. cerevisiae* for 217 experimental conditions is available (Aach et al., 2000; `http://arep.med.harvard.edu/ExpressDB`).

Groups Derived from Conserved Operons

Operons (tandem genes, usually functionally related, that are transcribed onto a single mRNA) are the main transcriptional regulatory units in bacteria. Often several bacterial operons that share an upstream binding site for a common transcription factor are coregulated to form a higher-order regulatory unit (a regulon). In different microbial organisms the genes that make up a regulon are often assorted into operons in different combinations. Two genes that are spatially separated on the chromosome in one organism, but spatially close in several other microbial species, are good candidates for being coregulated or functionally related (Dandekar et al., 1998; Overbeek et al., 1998, 1999). Such pairings between genes have been used to construct larger groups of genes that are predicted to be functionally coupled (Overbeek et al., 1999). Thus regulons can often be reconstructed by looking at conserved operon composition across other genomes. Figure 6.3 shows a simple example illustrating this method.

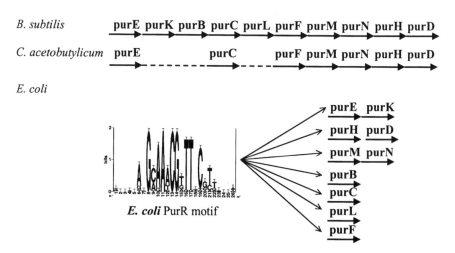

Figure 6.3 Predicting the *E. coli* purine biosynthesis regulon based on conserved operons. In *B. subtilis*, *C. acetobutylicum*, and several other bacteria, a large group of genes involved in purine biosynthesis is found together in a single operon. In *E. coli* these genes are split into seven different operons that are coordinately regulated by binding of the transcription factor PurR to a specific DNA motif found in the upstream region of each operon. Each color (not shown here) represents a separate but coregulated operon in *E. coli*.

A. M. McGuire and G. M. Church

In eukaryotes operon structures are less common, and each individual gene typically has an instance of the upstream regulatory motif in order to achieve coordinated transcription. However, two genes involved in the same pathway are often divergently transcribed from the same intergenic region and share an upstream regulatory site. Hence, divergently transcribed genes can be thought of as operons in a broader sense (Zhang and Smith, 1998).

Several of the groups constructed in this manner (Overbeek et al., 1999; McGuire and Church, 2000) correspond to known metabolic pathways, including the purine and arginine biosynthesis pathways in *E. coli*. By aligning the upstream regions of these groups of genes with a motif-finding algorithm, the regulatory motifs for the DNA-binding proteins that regulate these pathways (PurR and ArgR, respectively) can be found (McGuire et al., 2000; McGuire and Church, 2000). Therefore, the purine and arginine biosynthesis regulons, as well as the binding motifs for PurR and ArgR, can be found "from scratch," with no input of biological data other than the genome sequence itself. This method will become increasingly powerful as the number of genomes sequenced increases.

Groups Derived from Protein Fusions

Two genes found to be part of the same multidomain protein fusion in one organism, but scattered in the genome in another organism, are good candidates for being functionally related (Marcotte et al., 1999a, 1999b; Enright et al., 1999). Figure 6.4 shows a simple example illustrating this method. This method will become more powerful as the amount of genome sequence information increases.

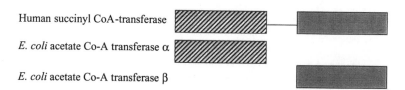

Figure 6.4 Illustration of the protein fusion method (Marcotte et al., 1999a, 1999b) of predicting functionally related genes. *E. coli* acetate Co-A transferases α and β are predicted to interact because their human homologues are fused into a single polypeptide chain. Boxes represent protein domains.

Groups Based on Protein Phylogenetic Profiles

Pellegrini et al. (1999) describe a method for determining functionally related groups of genes based on the assumption that proteins that function together in the cell are likely to evolve in a correlated fashion. In each species functionally linked genes (i.e., all of the proteins that make up a certain pathway) tend to be either all preserved or all eliminated. Thus, genes present in the same set of organisms and absent in the same set of organisms are likely to be functionally linked. Pellegrini et al. constructed groups of evolutionarily correlated genes by constructing a phylogenetic profile for each protein (a vector representing whether an orthologue to this protein is present or not present in each organism) and clustering these vectors based on their similarity. Figure 6.5 illustrates this method for predicting functional interactions between genes.

Groups Derived from Metabolic and Functional Pathways

Metabolic and functional group categories tabulated in existing databases are also useful data sets for motif finding. Some examples of such databases are KEGG (Kyoto Encyclopedia of Genes and Genomes; Ogata et al., 1999), MIPS (Munich Information Center for Protein Se-

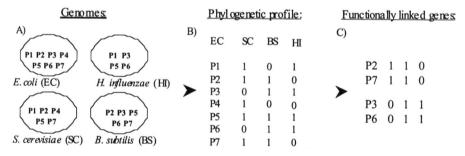

Figure 6.5 Predicting functionally related genes based on conserved evolution (Pellegrini et al., 1998). Panel (A) illustrates which of the seven *E. coli* proteins (P1–P7) are found in each of three additional bacterial genomes. In panel (B), a 1 or a 0 indicates that the protein is found or not found in this organism, respectively. The rows in panel (B) constitute the phylogenetic profile for this protein. By grouping genes with similar phylogenetic profiles, we obtain functionally linked groups of genes, illustrated in panel (C). P2 and P7 are predicted to be functionally linked, as are P3 and P6.

A. M. McGuire and G. M. Church

quences; Heinemeyer et al., 1998), and YPD (yeast protein database; Hodges et al., 1999). Since these groups are based on a compilation of experimental data in these organisms, the motifs found within these groups are biased toward known motifs. However, new motifs have also been found in the upstream regions of these groups. Hughes et al. (2000) used the MIPS and YPD groups to search for motifs in *S. cerevisiae*, and McGuire et al. (2000) used the KEGG groups to search for motifs in 17 complete prokaryotic genomes.

Groups Derived from Homologues to Footprinted Regulons in Other Species

Groups based on known regulons in other organisms, such as orthologues to the *E. coli* genes known to be regulated by 55 DNA-binding proteins (http://arep.med.harvard.edu/ecoli_matrices/; Robison et al., 1998), can also be used to search for upstream regulatory motifs. Such methods have been used to identify known regulatory motifs that are conserved in other organisms, as well as different motifs that could regulate orthologous sets of genes in different organisms (McGuire et al., 2000).

Combining Several Methods

Information obtained from a combination of the above methods can be used to obtain better predicted regulons. For example, McGuire and Church (2000) combined the three methods described above that rely only on information derived from comparative genomics (methods based on conserved operons, protein fusions, and phylogenetic profiles) to predict regulons in 24 complete genomes. These regulons were then used to search for new regulatory motifs.

MOTIF-FINDING ALGORITHMS

Once a group of genes believed to be coregulated (a predicted regulon) has been obtained by one of the methods described above, DNA regulatory motifs can be discovered by local alignment of the upstream regions of this set of genes (Roth et al., 1998). Several different algorithms have been used for motif discovery within DNA sequences. Only a sampling of the large number of motif-finding algorithms available will be mentioned here.

The Gibbs sampling algorithm is based on the statistical method of iterative sampling (Liu et al., 1995; Lawrence et al., 1993). This algorithm has been optimized for use with DNA sequence alignments in the program AlignACE (Roth et al., 1998; http://atlas.med.harvard.edu). The AlignACE algorithm, along with applications of AlignACE in bacterial genomes and in *S. cerevisiae*, will be discussed in more detail below.

MEME (Multiple EM for Motif Elicitation) is a motif-finding algorithm that uses an expectation maximization algorithm to locate one or more repeated patterns in the input sequence (Bailey and Elkan, 1995; Grundy et al., 1996; http://meme.sdsc.edu/meme/website/). MEME searches by building statistical models of motifs (matrices of discrete probability distributions) and maximizing the posterior probability of these models, given the data. MACAW (Multiple Alignment Construction and Analysis Workbench) is a motif-finding algorithm that allows the user to construct multiple alignments by locating, analyzing, editing, and combining blocks of aligned sequence segments (Schuler et al., 1991). Several methods based on detection of over-represented oligonucleotide frequencies have been proposed, including a method by van Helden et al. (1998; http://embnet.cifn.unam.mx/~jvanheld/rsa-tools). This simple and fast motif-finding method defines the statistical significance of a site based on tables of oligonucleotide frequencies in noncoding regions. However, this method is limited to short, highly conserved motifs.

AlignACE Algorithm

For clarity in describing the Gibbs sampling algorithm used in Align-ACE, we will present the simple case of searching for a single pattern of fixed width W within each of N input sequences ($S_1 \ldots S_N$). However, the AlignACE algorithm allows for automatic detection of variable pattern widths as well as multiple motif instances per input sequence. The pattern is described by a probabilistic model of residue frequencies at each position, q_{ij}, where i ranges from 1 to W and j ranges from 1 to 4 ($A, C, T,$ and G). We also consider the background frequencies for each base, p_j, where j ranges from 1 to 4. The p_{ij} are different for each or-

A. M. McGuire and G. M. Church

ganism. The start site for the motif within each input sequence is designated by a set of positions a_k, where k ranges from 1 to N. To choose the best pattern, the algorithm must choose values of a_k that maximize the ratio of the pattern probability to the background probability.

The first step is to choose random start positions (a_k). The next step (the predictive update step) is to choose one of the N sequences at random (z), and calculate the q_{ij} for the current positions, excluding z. The q_{ij} are calculated using the equation

$$q_{ij} = \frac{c_{ij} + b_j}{N - 1 + B}, \tag{6.1}$$

where c_{ij} is the count of nucleotide j at position i, b_j is a residue-dependent "pseudocount" suggested by Bayesian statistical analysis for the purpose of pattern estimation, and B is the sum of the B_j.

The next step (the sampling step) is to consider every possible segment x of length W within z, and to calculate the probability Q_x of generating each such segment x according to the current pattern probabilities q_{ij}, and the probability P_x of generating each segment according to the background probabilities p_j. A weight equal to $A_x = Q_x/P_x$ is assigned to segment x, and with each segment so weighted, a random position is selected. This new position becomes the new value for a_z, and the algorithm is iterated. The most probable alignment is the one that maximizes the equation

$$F = \sum_{i=1}^{W} \sum_{j=1}^{4} c_{ij} \log \frac{q_{ij}}{p_j}, \tag{6.2}$$

where c_{ij} is the count of nucleotide j at position i. As the pattern description calculated in the predictive update step becomes more accurate, the determination of the location in the sampling step also becomes more accurate.

Several optimizations of the Gibbs sampler algorithm were added in the AlignACE program for use in finding DNA motifs (Roth et al., 1998; Hughes et al., 2000): both strands of DNA are now considered; the near-optimum sampling method was improved in order to obtain higher scoring alignments; simultaneous multiple motif searching was replaced with an iterative masking approach; and the model for base background frequencies was set to equal the background nucleotide frequencies in the genome being considered.

INDICES FOR EVALUATING THE SIGNIFICANCE OF A MOTIF

When using a motif discovery algorithm such as AlignACE, often a large number of motifs are identified. Additional indices are necessary to assess these motifs and identify those most likely to correspond to specific regulatory motifs that are biologically relevant. Highly conserved and common genomic elements, such as repetitive elements and ribosome binding sites, are also found by AlignACE and must be separated from potential binding sites for specific transcription factors. Several indices used in assessing newly discovered motifs are described below.

MAP Score

The MAP score is the index used by AlignACE to determine the statistical significance of alignments sampled, given the composition of the input sequence (Liu et al., 1995). MAP scores are normalized so that the score for an alignment of zero sites is assigned a score of 0. The MAP score is higher for similar motifs with greater numbers of aligned sites and for more tightly conserved motifs, and lower for an identical alignment of sites derived from a larger set of input sequences, motifs with more dispersed information content, and motifs rich in nucleotides more prevalent in the genome. The MAP score is useful for ruling out alignments that are not statistically significant, but this score is not very useful in discriminating real regulatory motifs from other statistically significant motifs found by AlignACE because many nonspecific chromosomal features have high MAP scores (i.e., ribosome binding sites and repetitive elements).

Specificity Score

The specificity score is a measure of how specific a motif is for the sequence in which it was aligned, compared to the genome as a whole. This is the most useful index for selecting functional motifs. Two different specificity score measures have been described: the specificity score (S) is more useful for yeast and eukaryotes (Hughes et al., 2000), while the site specificity score (S_{site}) is more useful for bacterial sequences and genomes containing operons and multiple copies of a motif within a single upstream region (McGuire et al., 2000). The main difference between these two specificity scores is that S is a measure

A. M. McGuire and G. M. Church

of how many of the top predicted ORF targets in the genome were used to align the motif originally, whereas S_{site} measures the probability of obtaining the observed fraction of the top motif instances in the genome within the sequence input to AlignACE.

For each motif a position-specific weight matrix is constructed and the whole genome is searched for additional instances of the motif (Hughes et al., 2000; McGuire et al., 2000). The Berg and von Hippel weight matrix (Berg and von Hippel, 1987), implemented in the Scan-ACE program (Hughes et al., 2000), is used to perform this search. To calculate S_{site}, the top 200 ScanACE hits are considered, and the number of these sites located within the upstream regions used to align the motif is calculated. The probability of obtaining this number of sites within this subset of the genomic sequence by random means is calculated using the following equation (the hypergeometric distribution):

$$
S_{site} = \sum_{i=x}^{\min(s_1, s_2)} \frac{\binom{s_1}{i}\binom{N - s_1}{s_2 - i}}{\binom{N}{s_2}}.
\tag{6.3}
$$

N is the total number of possible sites (equal to the number of base pairs in the genome), s_1 is the number of ScanACE hits considered (200), s_2 is the total number of possible sites in the set of upstream regions used to align the motif (equal to the number of base pairs in the sequence input to AlignACE), and i is the intersection of s_1 and s_2. To calculate S, the 100 ORFs that are the most likely targets for the factor binding the motif are selected (the ORFs with the strongest sites in the region between -100 and 500 bp upstream of the translational start). This list of gene targets is compared against the list of genes in the group used to find the motif, using the same equation as above; however, now N is the total number of ORFs in the genome, s_1 and s_2 are the numbers of ORFs in the group used to find the motif and in the list of target genes, respectively, and x is the number of ORFs in the intersection of the two lists.

Positional Bias

This is a measure of bias in the locations of the top ScanACE hits in the genome relative to the start codon of the closest gene. The subset of the top 200 ScanACE hits that fall within a certain distance s of the start codon (300 bp for bacteria and 600 bp for *S. cerevisiae*) are considered. A smaller window w (30 bp for bacteria and 50 bp for *S. cerevisiae*) is then chosen such that it contains the maximum number of hits within s. The probability of seeing the observed number of sites or greater in a window of size w, out of a sample space of size s, is calculated as follows (Hughes et al., 2000):

$$P = \sum_{i=m}^{t} \binom{s_1}{i} \left(\frac{w}{s}\right)^i \left(1 - \frac{w}{s}\right)^{t-i} \tag{6.4}$$

Similarities Between Motifs

The CompareACE program (Hughes et al., 2000), used to compare two motifs by finding the best alignment, calculates the Pearson correlation between position-specific scoring matrices (Pietrokovski, 1996). Two motifs with a CompareACE score greater than 0.7 are considered similar (Hughes et al., 2000; McGuire et al., 2000). New motifs can be compared against databases of known motifs using CompareACE, including a database of footprinted *E. coli* motifs (Robison et al., 1998). CompareACE scores can also be used as a metric to cluster motifs. After calculating a matrix of pairwise CompareACE scores, motifs can be clustered using a simple joining algorithm (Hartigan, 1975). Often when a large set of motifs is analyzed, multiple instances of similar motifs will be present, and clustering will reduce the set of motifs to unique clusters.

Palindromicity

Many known bacterial motifs are palindromic. To identify palindromic motifs, the CompareACE program can be used to compare a motif and its reverse complement. The same CompareACE cutoff score as for comparing motifs to one other (0.7) has been used (McGuire et al., 2000).

ALIGNACE APPLICATIONS

Applications of AlignACE in Complete Bacterial Genomes

AlignACE has been applied to several kinds of potentially coregulated groups of genes in prokaryotic genomes. McGuire et al. (2000) used groups derived from conserved operons in the WIT database, groups derived from metabolic pathways in the KEGG database, and groups derived from homologues in other species to members of known *E. coli* regulons (Robison et al., 1998) to search for regulatory motifs in 17 prokaryotic genomes. Specific, new motifs were found using each grouping method and in each organism. McGuire and Church (2000) combined and compared three techniques for predicting functional interactions based on comparative genomics (methods based on conserved operons, protein fusions, and correlated evolution) in order to construct predicted regulons in 22 complete prokaryotic genomes. These predicted regulons were then used to find a large number of new motifs using AlignACE.

Identifying motifs in bacterial genomes is complicated by the presence of operons. It is difficult to locate the regulatory region for a gene within an operon. If the gene of interest lies within an operon, its promoter could lie several genes upstream. Thus, we must include several possible intergenic regions to ensure that we have included the segment containing the regulatory region (McGuire et al., 2000). This is possible in bacteria because regulatory regions are shorter; hence, we can include several intergenic regions without adding so much extra sequence that the motif-finding algorithm will no longer be able to align the regulatory motif.

An additional problem associated with motif finding in bacteria is that there are fewer instances of most regulatory motifs because there is usually only one instance per operon instead of one instance per gene. Motif conservation can used to aid in finding new motifs by grouping upstream regions from closely related bacteria, thus increasing the number of instances of each motif in the sequence to be aligned (McGuire et al., 2000).

The effectiveness of AlignACE in finding known *E. coli* motifs was analyzed. Of the 32 *E. coli* footprinted regulons in our database with 5–100 known binding sites (Robison et al., 1998), 26 have motifs that can be found by AlignACE (81%). The groups obtained from homologues in other species to members of known *E. coli* regulons were used to study conservation of *E. coli* DNA motifs in other organisms, as well as to identify potential new mechanisms for regulating the same cellular process in more distantly related organisms. At least 30% of the known *E. coli* DNA regulatory motifs were found to be conserved in closely related bacteria. A new motif can indicate either a different mechanism for regulating a similar cellular process or divergence of binding site residues in a conserved DNA-binding protein.

Applications of AlignACE in *Saccharomyces cerevisiae*

AlignACE has been used for motif finding in *S. cerevisiae* with mRNA expression data (Roth et al., 1998; Tavazoie et al., 1999), as well as with functional group categories (Hughes et al., 2000). Selecting upstream regions is more straightforward in yeast because there are no operons. Because intergenic regions are longer in yeast, and regulatory elements are often found farther upstream than in bacteria, longer stretches of noncoding sequence have been used for motif finding (up to 600 bp).

Roth et al. (1998) first used AlignACE to find motifs, employing regulons derived from experiments employing Affymetrix GeneChips in the following three systems: galactose utilization, mating type, and heat shock response. By aligning the groups of genes with the largest fold changes in these conditions, using AlignACE, they were able to find the binding sites for many of the factors known to be involved in these processes.

Tavazoie et al. (1999) looked at a time series of mRNA abundance data measured over two synchronized *S. cerevisiae* cell cycles. These data were used to group 3000 yeast genes into 30 clusters with similar temporal expression profiles. These groups were highly enriched with genes involved in the same cellular processes. By aligning the genes making up these clusters with AlignACE, they were able to find many known motifs as well as several new ones.

Hughes et al. (2000) used AlignACE to search for motifs upstream of the genes making up the MIPS and YPD (yeast protein database) functional groups. Many known and new motifs were found in this analysis, including a new motif found upstream of genes involved in proteasome formation. This motif has been independently identified as the binding site for Rpn4 in Church's lab as well as by Mannhaupt et al. (1999). Several other potentially functional new *S. cerevisiae* motifs are being experimentally tested.

CONCLUSIONS

The use of motif-finding algorithms to discover new regulatory motifs is a powerful method for understanding the interconnections in genetic regulatory networks. Groups of genes (predicted regulons) obtained from mRNA expression data can be extremely powerful in this regard, though such groups of genes have the drawback that they are dependent on the experimental conditions chosen. These groups can be complemented by theoretically obtained predicted regulons that are less dependent on the experimental conditions chosen. The presence of a significant regulatory motif shared by the members of a predicted regulon provides additional evidence that this group of genes is actually coregulated.

The motif-finding program AlignACE has been used successfully to predict new motifs in *S. cerevisiae*, as well as in over 20 complete bacterial genomes. A new motif discovered by AlignACE upstream of proteasome subunits in *S. cerevisiae* has been confirmed experimentally, and numerous other new motifs should be verified as well. AlignACE has also been used to analyze conservation and divergence of regulatory mechanisms between microbial genomes.

The methods that we have described here for predicting regulons and their cis-regulatory motifs will become increasingly powerful as the size of the sequence database and the number of complete genomes grows. The discovery of DNA regulatory motifs will be facilitated by having complete genomes for a greater number of closely related organisms, since sequences from closely related organisms can be pooled to find conserved motifs that are found in only a few copies in each genome. In addition, regulon-prediction methods based on comparative genomics (including the methods described here that involve conserved operons, protein fusions, and phylogenetic profiles) will yield more information about each organism as more complete and partial genome sequences become available. Several complete genomes have recently been made available, including *Drosophila*, *C. elegans*, and several human chromosomes. As many as 50 additional bacterial genomes could become available by 2002, along with the human genome sequence. The framework described here for predicting regulons and their cis-regulatory motifs will soon become increasingly powerful.

ACKNOWLEDGMENTS

We would like to thank Jason Hughes, Jason Johnson, Martha Bulyk, Saeed Tavazoie, Pete Estep, John Aach, Jong Park, Adnan Derti, Tzachi Pilpel, and Jeremy Edwards for help and discussions, as well as all of the other members of George Church's lab. Abigail Manson McGuire is a Howard Hughes Medical Institute predoctoral fellow.

REFERENCES

Aach, J., Rindone, W., and Church, G. M. (2000). Systematic management and analysis of yeast gene expression data. *Genome Res.* 10: 431–445.

Bailey, T. L., and Elkan, C. (1995). The value of prior knowledge in discovering motifs with MEME. In *Proceedings of the Third International Conference in Intelligent Systems for Molecular Biology*. Cambridge, UK: AAAI Press, pp. 21–29.

Berg, O. G., and von Hippel, P. H. (1987). Selection of DNA binding sites by regulatory proteins. Statistical-mechanical theory and application to operators and promoters. *J. Mol. Biol.* 193: 723–750.

Dandekar, T., Snel, B., Huynen, M., and Bork, P. (1998). Conservation of gene order: A fingerprint of proteins that physically interact. *Trends Biochem. Sci.* 23(9): 324–328.

Eisen, M. B., Spellman, P. T., Brown, P. O., and Botstein, D. (1998). Cluster analysis and display of genome-wide expression patterns. *Proc. Natl. Acad. Sci. USA* 95: 14863–14868.

Enright, A. J., Iliopoulos, I., Kyrpides, N. C., and Ouzounis, C. A. (1999). Protein interaction maps for complete genomes based on gene fusion events. *Nature* 402: 86–90.

Grundy, W. N., Bailey, T. L., and Elkan, C. P. (1996). ParaMEME: A parallel implementation and a web interface for a DNA and protein motif discovery tool. *Comput. Appl. Biosci.* 12: 303–310.

Hartigan, J. A. (1975). *Clustering Algorithms*. New York: Wiley.

Heinemeyer, T., Wingender, E., Reuter, I., Hermjakob, H., Kel, A. E., Kel, O. V., Ignatieva, E. V., Ananko, E. A., Podkolodnaya, O. A., Kolpakov, F. A., Podkolodny, F. A., and Kolchanov, N. A. (1998). A database on transcription regulation: TRANSFAC, TRRD, and COMPEL. *Nucleic Acids Res.* 26: 364–370.

Hodges, P., McKee, A., Davis, B., Payne, W., and Garrels, J. (1999). Yeast protein database (YPD): A model for the organization and presentation of genome-wide functional data. *Nucleic Acids Res.* 27: 69–73.

Hughes, J. D., Estep, P. W., Tavazoie, S., and Church, G. M. (2000). Computational identification of cis-regulatory elements associated with groups of functionally related genes in *Saccharomyces cerevisiae*. *J. Mol. Biol.* 296: 1205–1214.

Lawrence, C. E., Altschul, S. F., Boguski, M. S., Liu, J. S., Neuwald, A. F., and Wootton, J. C. (1993). Detecting subtle sequence signals: A Gibbs sampling strategy for multiple alignment. *Science* 262: 208–214.

Liu, S. J., Neuwald, A. F., and Lawrence, C. E. (1995). Bayesian models for multiple local sequence alignment and Gibbs sampling strategies. *J. Am. Stat. Assoc.* 90: 1156–1170.

Mannhaupt, G., Schnall, R., Karpov, V., Vetter, I., and Feldmann, H. (1999). Rpn4 acts as a transcription factor by binding to PACE, a nonamer box found upstream of 26S proteasomal and other genes in yeast. *FEBS Lett.* 450: 27–34.

Marcotte, E. M., Pellegrini, M., Ng, H. L., Rice, D. W., Yeates, T. O., and Eisenberg, D. (1999). Detecting protein function and protein-protein interactions from genome sequences. *Science* 285: 751–753.

Marcotte, E. M., Pellegrini, M., Thompson, M. J., Yeates, T. O., and Eisenberg, D. (1999). A combined algorithm for genome-wide prediction of protein function. *Nature* 402: 83–86.

McGuire, A. M., and Church, G. M. (2000). Predicting regulons and their cis-regulatory motifs by comparative genomics. *Nucleic Acids Res.* 28(22): 4523–4530.

McGuire, A. M., Hughes, J. D., and Church, G. M. (2000). Conservation of DNA regulatory motifs and discovery of new motifs in microbial genomes. *Genome Res.* 10: 744–757.

Ogata, H., Goto, S., Sato, K., Fujibuchi, W., Bono, H., and Kanehisa, M. (1999). KEGG: Kyoto Encyclopedia of Genes and Genomes. *Nucleic Acids Res.* 27(1): 29–34.

Overbeek, R., Fonstein, M., D'Souza, M., Pusch, G., and Maltsev, N. (1998). *In silico biology, 1:0009.* http://www.bioinfo.de/isb/1998/01/0009.

Overbeek, R., Fonstein, M., D'Souza, M., Pusch, G. D., and Maltsev, N. (1999). The use of gene clusters to infer functional coupling. *Proc. Natl. Acad. Sci. USA* 96: 2896–2901.

Pellegrini, M., Marcotte, E. M., Thompson, M. J., Eisenberg, D., and Yeates, T. O. (1999). Assigning protein functions by comparative genome analysis: Protein phylogenetic profiles. *Proc. Natl. Acad. Sci. USA* 96: 4285–4288.

Pietrokovski, S. (1996). Searching databases of conserved sequence regions by aligning protein multiple-alignments. *Nucleic Acids Res.* 24: 3836–3845.

Robison, K., McGuire, A. M., and Church, G. M. (1998). Comprehensive library of DNA-binding site matrices for 55 proteins applied to the complete *Escherichia coli* K-12 genome. *J. Mol. Biol.* 284: 241–254.

Roth, F. P., Hughes, J. D., Estep, P. W., and Church, G. M. (1998). Finding DNA regulatory motifs within unaligned noncoding sequences clustered by whole-genome mRNA quantitation. *Nat. Biotechnol.* 16: 939–945.

Schneider, T. D., and Stephens, R. M. (1990). Sequence logos: A new way to display consensus sequence. *Nucleic Acids Res.* 18: 6097–6100.

Schuler, G. D., Altschul, S. F., and Lipman, D. J. (1991). A workbench for multiple alignment construction and analysis. *Proteins: Struct. Funct. Genet.* 9: 180–190.

Sherlock, G. (2000). Analysis of large-scale gene expression data. *Curr. Opinion Immunol.* 12: 201–205.

Tamayo, P., Slonim, D., Mesirov, J., Zhu, Q., Kitareewan, S., Dmitrovsky, E., Lander, E. S., and Golub, T. R. (1999). Interpreting patterns of gene expression with self-organizing maps: Methods and application to hematopoietic differentiation. *Proc. Natl. Acad. Sci. USA* 96: 2907–2912.

Tavazoie, S., Hughes, J. D., Campbell, M. J., Cho, R. J., and Church, G. M. (1999). Systematic determination of genetic network architecture. *Nat. Genetics* 22: 281–285.

Van Helden, J., André, B., and Collado-Vides, J. (1998). Extracting regulatory sites from the upstream region of yeast genes by computational analysis of oligonucleotide frequencies. *J. Mol. Biol.* 281: 827–842.

Wen, X., Fuhrman, S., Michaels, G. S., Carr, D. B., Smith, S., Barker, J. L., and Somogyi, R. (1998). Large-scale temporal gene expression mapping of central nervous system development. *Proc. Natl. Acad. Sci. USA* 95: 334–339.

Zhang, X., and Smith, T. F. (1998). Yeast "operons." *Microb. Comp. Genomics* 3: 133–140.

SUGGESTED READING

Brazma, A., Jonassen, I., Eidhammer, I., and Gilbert, D. (1998). Approaches to the automatic discovery of patterns in biosequences. *J. Comput. Biol.* 5(2): 279–305. Survey of some other types of pattern discovery algorithms.

Brown, P. O., and Botstein, D. (1999). Exploring the new world of the genome with DNA microarrays. *Nat. Genetics* 21: 33–37. Review on microarray technology.

Eisenberg, D., Marcotte, E. M., Xenarios, I., and Yeates, T. O. (2000). Protein function in the post-genomic era. *Nature* 405: 823–826. Review of methods for predicting interactions between nonhomologous genes.

Gralla, J. D., and Collado-Vides, J. (1996). Organization and function of transcription regulatory elements. In F. C. Neidhardt (ed.), *Escherichia coli and Salmonella: Cellular and Molecular Biology*, 2nd ed. Washington, D.C.: American Society for Microbiology, pp. 1232–1245. Review on bacterial transcriptional regulatory motifs.

Sherlock, G. (2000). Analysis of large-scale gene expression data. *Curr. Opinion Immunol.* 12: 201–205. Review of methods for cluster analysis of microarray data.

URLs FOR RELEVANT SITES

Church lab web page. `http://arep.med.harvard.edu`

Course Web page for IBSS.
`http://arep.med.harvard.edu/labgc/amcguire/IBSS/index.html`

Motif Discovery Software
GibbsDNA (another Gibbs sampling program available on the Web).
`http://argon.cshl.org/ioschikz/gibbsDNA/`

Home page for AlignACE; AlignACE and accessory programs can be downloaded here.
`http://atlas.med.harvard.edu/`

MEME (Multiple EM for Motif Elicitation).
`http://meme.sdsc.edu/meme/website/`

PRATT program for pattern discovery.
`http://www.ii.uib.no/~inge/Pratt.html`

Regulatory sequence analysis tools at UNAM.
`http://copan.cifn.unam.mx/~jvanheld/rsa-tools/`

Regulon Prediction

DNA binding-site matrices for 59 known *E. coli* DNA regulatory motifs.
`http://arep.med.harvard.edu/ecoli_matrices/`

ExpressDB RNA expression database.
`http://arep.med.harvard.edu/cgi-bin/ExpressDByeast/EXDStart`

KEGG (Kyoto Encyclopedia of Genes and Genomes) database (metabolic pathways and functionally related groups of genes). `http://star.scl.genome.ad.jp/kegg/`

WIT (What Is There?) database (contains database of conserved operons).
`http://wit.mcs.anl.gov/WIT2/`

Yeast protein function assignment, using phylogenetic profiles and domain fusions.
`http://www.doe-mbi.ucla.edu/people/marcotte/yeast.html`

7 Gene Networks Description and Modeling in the GeneNet System

Nikolay A. Kolchanov, Elena A. Ananko, Vitali A. Likhoshvai, Olga A. Podkolodnaya, Elena V. Ignatieva, Alexander V. Ratushny, and Yuri G. Matushkin

The molecular genetic systems that control the processes occurring in organisms on the basis of the hereditary information contained in their genomes are called gene networks.

Numerous biological, biochemical, and physiological molecular processes occur simultaneously in humans, animals, and plants, as well as in prokaryotes, eukaryotes, and archaea. Cells divide and differentiate, tissues and organs are formed. Organisms enter into complex interactions with the environment while consuming matter, energy, and information flows during their growth, development, and reproduction. All of these diverse processes are regulated genetically. Gene networks, with groups of genes functioning in concert as their central elements, are the backbone of this regulation.

Theoretical studies of gene networks commenced in the 1960s. They considered general organization patterns of molecular genetic systems controlling functions of prokaryotes (Ratner, 1966) and described the dynamics of gene networks within the simplest logic schemes (Kauffman, 1969; Sugita, 1961). Further studies involved approaches based on differential equations (Thomas, 1973; Savageau, 1985; Bazhan et al., 1995; Belova et al., 1995; Likhoshvai et al., 2000) and stochastic models (McAdams and Arkin, 1997). Numerous methods for mathematical simulation have been developed, including (1) Boolean networks (Sanchez et al., 1997) allowing gene networks to be reconstructed from experimental data (Wahde and Hertz, 2000); (2) the logical approach (Thieffry and Thomas, 1995; Sanchez et al., 1997); (3) Petri nets (Hofestadt and Meineke, 1995); and (4) threshold models (Tchuraev, 1991).

However, insufficient experimental data limited the development of the theory of gene networks until the mid-1980s. During the 1990s, the appearance of efficient methods for studying molecular mechanisms that regulate gene expression and successful research in structural-functional organizations of various genomes triggered an explosive accumulation of experimental data on functions of gene networks. This accumulation led to a wide diversity of databases on various features of the gene network functions, genetic regulation of metabolic processes, morphogenesis, development, and so on. It was a powerful stimulus for developing both new and earlier methods, approaches, and algorithms designed to detect functional regularities of biological systems at all levels of organization.

The search for regularities based on contextual analysis of nucleotide sequences is considered by McGuire and Church in chapter 6 of this volume. In chapter 8 of this volume, Huang introduces the application of logical approaches to the development of multicellular organisms. In this chapter we consider the problems connected with regularities of gene network function and point out two aspects of the question: (1) developing methods for storing and preserving the information accumulated both in experiments and by means of numeric analysis of mathematical models; (2) working out the methods of compiling and analyzing mathematical models of gene network functioning.

The databases listed here and many other databases are an important source of information for both the experimental study and the computer analysis of gene networks, genetically controlled metabolic processes, physiological systems, and other items. The role and significance of such databases will grow with the amount of experimental information on functions of gene networks and genetically controlled systems and processes. Consequently development of efficient technologies for accumulating this information in computer databases is of the utmost importance.

Detailed here is the technology we have developed for computer-assisted description of gene networks and the database GeneNet, constructed using this technology. Analysis of a variety of actual gene networks (Ananko et al., 1997; Ignatieva et al., 1997; Podkolodnaya and Stepanenko, 1997; Merkulova et al., 1997) suggested us two important methodological principles for development of the technology in question:

1. The function of any gene network in either a unicellular or a multicellular organism involves a limited set of elementary structures and

events of their interactions at different hierarchical levels of organization (genes, cell nucleus, cytoplasm, nuclear membrane, intercellular space, tissue, or organ). The specific combination of elementary units and events generates a tremendous diversity of gene networks with typical patterns of structural-functional organization and functional modes.

2. In any genetically controlled organism system, it is impossible to separate in a pure form the genetic component itself (i.e., the component that performs the control) and the controlled component (which provides for a particular biochemical, physiological, or other elementary function). This means that description of a gene network implies simultaneous consideration of these components.

This chapter details the technology of gene network description; the GeneNet database (`http://wwwmgs.bionet.nsc.ru/systems/mgl/genenet/`), created using this technology and containing the information on more than 20 gene networks of multicellular organisms; and basic principles of the gene network organization and functions. We also describe mathematical simulation of biological systems and illustrate the method used by particular models of gene network dynamics.

TECHNOLOGY FOR DESCRIBING GENE NETWORKS

Object-Oriented Approach

In the object-oriented approach (Schweigert et al., 1995) the components of gene networks in the GeneNet database are divided into elementary structures (entities) and elementary events (relationships between the entities). The hierarchy of GeneNet object classes is shown in figure 7.1.

Elementary Structures We consider the following elementary structures to be significant for the function of gene networks: gene, RNA, protein, and nonproteinaceous substance (figure 7.1). This list can be extended, if necessary, with other elementary structures.

Each object class is described in a separate table, using a specialized data representation format that takes into account the peculiarities of each class. The database contains the following tables: GENE, RNA, PROTEIN, and SUBSTANCE.

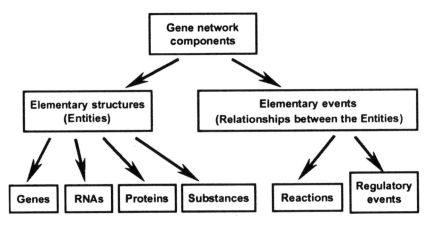

Figure 7.1 Class hierarchy in the GeneNet database.

Elementary Events In the GeneNet database we consider two types of elementary events: reaction and regulatory event (figure 7.1).

Chemical notation is the basis for describing the elementary events. Thus, any event is described as follows:

$$C_1, \ldots, C_k$$
$$\downarrow$$
$$A_1 + \cdots A_n \Rightarrow B_1 + \cdots B_m,$$

where A's are the elementary entities in the interaction; C's are the entities modulating this interaction; and B's are the products of the interaction. Based on this model, we consider two types of interactions between gene network components:

1. Reaction, the interaction between entities that leads to generation of a new entity (assembly or disassembly of a multimeric complex, expression of a protein, secretion of certain substances, protein modifications, etc.). In some cases such a reaction corresponds to a single biochemical reaction (e.g., protein phosphorylation), while in other cases it corresponds to a series (cascade) of successive biochemical reactions (e.g., expression of a protein). Interactions of the former type are designated in the GeneNet database as direct reactions; of the latter type, as indirect reactions.

2. Regulatory event, the effect of an entity (it may be a catalyst or an inhibitor) on a certain reaction. Regulatory events of four types are

distinguished, depending on their effect on the reaction: switching on, switching off, positive effect, and negative effect. Regulatory events described in the GeneNet database are induction of gene expression, repression of gene expression, activation of signal transduction pathway by effector molecules, enzymatic catalysis etc.

Levels of Gene Network Representation

The GeneNet system allows the fact that gene network components may be distributed in various organs, tissues, cells, and cell compartments to be taken into account. As a first approximation, three main hierarchical levels (organism, cell, and gene) are considered in the description of the gene network.

Organism Level The entities described at this level are organs, tissues, particular types of cells, and secreted proteins and substances affecting other organs, tissues, and cells. The description at this level enables the spatial order of gene network components in the organism to be considered.

Cell Level The entities described at this level are cell compartments (for example, cytoplasm, nucleus, and mitochondria), proteins, RNAs, genes, and substances (e.g., steroids, lipids, energy-stored molecules, metabolites). The description at this level enables distribution of the gene network components throughout cell compartments to be considered.

Gene Level Regulation of gene expression is described in detail at this level employing the information from the TRRD database (Kolchanov et al., 2000).

FORMAT FOR DESCRIBING ELEMENTARY STRUCTURES AND ELEMENTARY EVENTS

Elementary Structures

The elementary structures essential for the function of gene networks in the GeneNet database are described in the tables GENE, PROTEIN,

```
ID   Gg:GATA1-p
DT   09.10.97; Ananko E.; created.
DT   19.6.1999.; Podkolodnaya O.A.; updated.
OS   Gallus gallus (chicken).
SN   GATA1
NM   transcription factor GATA1
SY   NF-E1
FN   active
MM   no data
MD   phosphorylated
GN   Gg:GATA-1
DR   TFFACTOR; T00267;
DR   SWISSPROT; P17678;
DR   EMBL; M26209;
DR   PIR; A32993;
SO   Gg:Eryth Prog
RF   Briegel K., et al. 1996
//
```

Figure 7.2 Description of the elementary structure in the table PROTEIN of GeneNet database, exemplified by GATA1 transcription factor, which is essential for the function of the gene network of erythrocyte differentiation and maturation.

RNA, and SUBSTANCE. The data format used for describing the elementary structures involves several information fields.

For example, the entry shown in figure 7.2 indicates that chicken protein GATA1-p is a transcription factor (NM) that is phosphorylated (MD); exists in an active state (FN); has a synonymous name NF-E1 (SY); is described in databases TFFACTOR, SWISS-PROT, EMBL, and PIR (DR); and is isolated from erythroid progenitor cells (SO); and the information stored was obtained from a paper by K. Briegel et al. (1996) (RF).

Elementary Events

Two types of interactions are considered when describing the gene networks in the GeneNet database: reactions and regulatory events (see above). In the GeneNet database, elementary events are described in the table RELATION.

a)
```
ID  <protein>Hs:preSREBP1^cytoplasm ->
<protein>Hs:SREBP1^cytoplasm
DT  02.8.1999.; Ignatieva E.V.; created.
EF  direct
RF  Wang X. et al., 1994
//
```

b)
```
ID  <substance>Cholesterol^cytoplasm ->>
<protein>Hs:SRP^cytoplasm ->> <protein>Hs:preSREBP1^cytoplasm
-> <protein>Hs:SREBP1^cytoplasm
DT  02.8.1999.; Ignatieva E.V.; created.
AT  decrease
EF  indirect
RF  Wang X. et al., 1994
//
```

Figure 7.3 Description of relationships in the GeneNet database. (a) reaction—transformation of the inactive precursor transcription factor preSREBP1 into the mature factor SREBP1; (b) regulatory event: inhibition of SRP protease activity by cholesterol (gene network of lipid metabolism).

Reactions The data format used for describing reactions involves several information fields.

Description of a reaction within the gene network of lipid metabolism, shown in figure 7.3a, includes transformation of transcription factor preSREBP1 into active transcription factor SREBP1, the key regulator of this gene network. The record means that the human protein preSREBP1 transforms into human protein SREBP1, the reaction takes place in the cytoplasm (ID), and the interaction described is direct (EF).

As is described below, this reaction requires sterol-regulated protease (SRP). The activity of this protease decreases with the increase in cell cholesterol level, thus closing the negative feedback circuit controlling the cholesterol level in the cell.

Description of a regulatory event significant for the function of the gene network of lipid metabolism (i.e. inhibition of SRP by cholesterol) is shown in figure 7.3b. The SRP protease provides for activation of the inert transcription factor preSREBP1 into the operative SREBP1. The record shown means that cholesterol inhibits (AT) the reaction transforming preSREBP1 into active SREBP1, which requires SRP (ID). The reaction takes place in the cytoplasm and is described as indirect (EF).

DATA INPUT AND VISUALIZATION OF GENE NETWORK STRUCTURE

Since the language described is rather complex, the user deals only with a specially developed graphic interface (GeneNet Data Input GUI) that generates and interprets the code of the language (Kolpakov and Ananko, 1999).

With the help of this program, the user can input the data into the GeneNet database, operating with concepts of molecular biology associated with the expression of genes. The input interface automatically translates the input information into a standard GeneNet format, examples of which were considered above.

A specialized JAVA program, GeneNet Viewer, processes the formalized data accumulated in the GeneNet database and presents it to the user as a graphic diagram (Kolpakov et al., 1998). The Viewer allows the GeneNet database to be explored and visualized through the Internet and includes tools for automated generation of gene network diagrams, a system of filters, tools for data navigation, on-line help, interactive cross-references within the GeneNet database, and references to other databases (figure 7.4). A standard set of images corresponding to elementary structures and events is used for gene network visualization. Figure 7.5 demonstrates standard GeneNet graphical representation of the elementary structures and events, some of which (figure 7.5c,d) are described in figure 7.3.

TYPICAL EXAMPLES OF GENE NETWORKS DESCRIBED IN THE GENENET DATABASE

Analysis of the information contained in the GeneNet database and the available literature suggests several basic types of gene networks.

1. Gene networks controlling cyclic processes, such as the cell cycle and the cycle of heart muscle contraction

2. Gene networks underlying cell growth and differentiation, morphogenesis of tissues and organs, growth and development of the organism

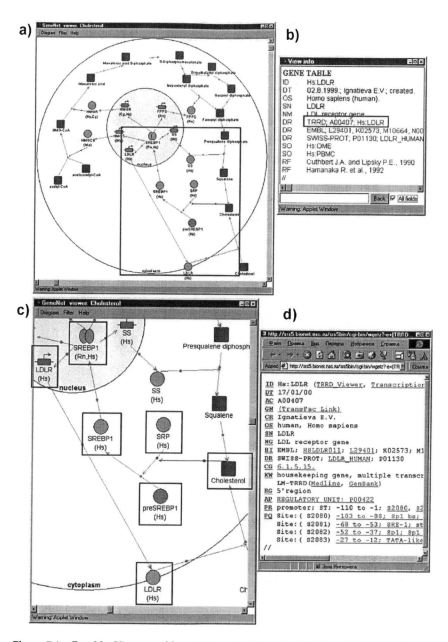

Figure 7.4 GeneNet Viewer and hypertext navigation: visualization of the gene network of lipid metabolism. Inner circle, nucleus; outer circle, cytoplasm enclosed by cell membrane; black rectangles, genes; and small gray circles, protein molecules. Black squares indicate nonproteinaceous compounds of the cholesterol biosynthesis pathway. (a) Gene network of lipid metabolism. (b) Text window with the human LDLR gene description from the GeneNet database. (c) Zooming of the diagram. Elementary structures (genes, proteins, and metabolites) involved in the regulation of this gene network according to the negative feedback mechanism (cholesterol, proteins preSREBP1, SREBP1, SRP, LDRL, and gene *LDLR*) are framed. (d) Hypertext reference to the TRRD database.

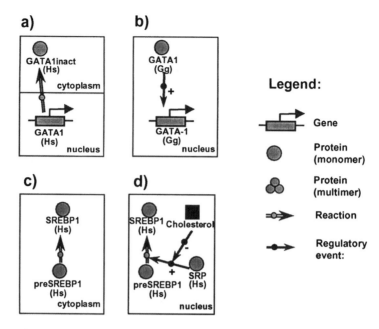

Figure 7.5 Graphical representation of elementary structures and events essential for the gene network function. (a) The GATA1 gene expression resulting in the emergence of the inactive precursor of the transcription factor GATA1inact in the cytoplasm. (b) Autoactivation of GATA-1 gene transcription by factor GATA1 encoded by this gene (gene network of erythrocyte differentiation and maturation). (c) Transformation of the transcription factor preSREBP1 inactive precursor into the mature factor SREBP1. (d) Inhibition of SRP protease activity by cholesterol (gene network of lipid metabolism).

3. Gene networks maintaining homeostases of biochemical and physiological parameters of the organism

4. Gene networks providing for responses of the organism to changes in the environment (e.g., stress response).

We will now consider certain typical patterns of gene networks.

Gene Networks Maintaining Homeostasis of the Organism

Negative feedback regulation plays an important role in the operation of gene networks, providing for maintenance of a parameter within a certain range around its optimal level (figure 7.6a). An example is the gene network regulating intracellular cholesterol concentration (figure

N. A. Kolchanov et al.

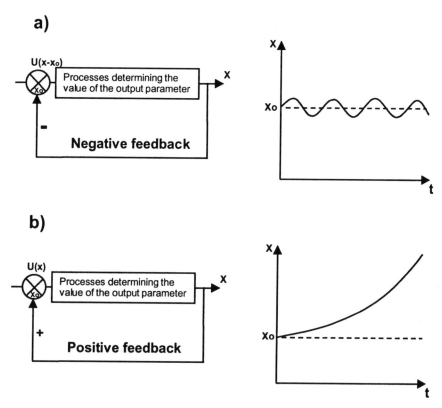

Figure 7.6 Two basic types of regulatory circuits in the gene networks. (a) Negative feedback. (b) Positive feedback.

7.4a). Cholesterol synthesis is implemented with the involvement of mevalonate pathway enzymes (Ericsson et al., 1996). Increased transcription of the genes coding for these enzymes raises the intracellular cholesterol concentration.

The key regulators of this pathway are transcription factors of the SREBP subfamily. They activate the transcription of a cassette of genes coding for many enzymes of the mevalonate pathway, since regulatory regions of all these genes contain binding sites for these factors.

SREBP1 is formed through proteolytic cleavage of its inactive precursor (preSREBP1), with a molecular weight of 125 kDa, performed by a sterol-regulated protease, resulting in the active factor, with a molecular weight of 68 kDa (Wang et al., 1995).

An increase in cholesterol level suppresses SRP activity, slowing the transition of preSREBP1 into the active form and decreasing the level of active SREBP1 and the transcriptional activities of the genes of the mevalonate pathway. In turn, decreases in the levels of the mevalonate pathway enzymes reduce the rate of cholesterol synthesis, thus normalizing its level in the cell.

The intensity of cholesterol transport across the cell membrane from the intercellular space plays an important role in the maintenance of the intracellular level of cholesterol. The transport involves low-density lipoprotein receptor (LDLR). At a decreased cholesterol concentration, the concentration of active SREBP1 is increased, activating transcription of the gene encoding LDLR (Lloyd and Thompson, 1995). With an increased intracellular cholesterol level, the activities of sterol-regulated proteases, and therefore the concentration of active SREBP, decrease. This in turn decreases the transcription activity of the LDLR gene and cholesterol transport into the cell (figure 7.4a, c).

A characteristic feature of the gene network being considered is its activation by the decrease in the level of the parameter it adjusts (cholesterol concentration), and its halting when the level of this parameter exceeds the optimal value. Such mechanisms are present in virtually all the homeostatic gene networks (e.g., gene networks controlling the thyroid system and physiological redox homeostasis, which are also described in the GeneNet database).

Gene Networks Controlling Irreversible Processes

Cell differentiation, the morphogenesis of tissues and organs, and the growth and development of organisms are examples of gene network-controlled irreversible processes. Typically the gene networks controlling irreversible processes (1) are triggered by an external signal and (2) contain positive feedback circuits, which essentially boost the external signal and thereby trigger the irreversible processes of the gene network function. Such circuits provide for the maximal effective deviation of a parameter from its current value (figure 7.6b).

An example of the gene network providing for differentiation and maturation of erythrocytes is shown in figure 7.7. The key external regulator of this gene network is erythropoietin (EPO). Binding of erythropoietin to the membrane receptor (EPOR) results in dimerization of the latter (Elliott et al., 1996), triggering the signal transduction

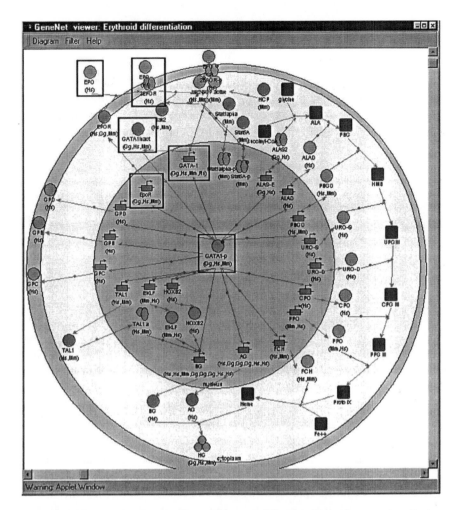

Figure 7.7 Gene network controlling erythrocyte differentiation and maturation. Designations as in figure 7.5; black squares indicate initial, intermediate, and final compounds of the heme biosynthesis pathway. Elementary structures (genes and proteins) activating this gene network through positive feedback mechanisms (EPO, protein complex EPO/EPOR, transcription factor GATA1-p, and genes *GATA-1* and *EpoR*) are framed.

pathway implemented by cell protein kinases. As a result, certain transcription factors are phosphorylated, transferred to the nucleus, and acetylated (Boyes et al., 1998; Zhang et al., 1998). They activate transcription of several genes, including the gene coding for transcription factor GATA1 (figure 7.7) (Dalyot et al., 1993).

Transcription factor GATA1 is the key regulator of erythrocyte differentiation and maturation. The presence of GATA1 binding sites in regulatory regions of virtually all the genes constituting this gene network underlies its key regulatory role.

The transcription factor GATA1 provides two positive feedback circuits that boost the initial differentiation signal received by the erythropoietin receptor through its interaction with erythropoietin, and conveyed to the cell nucleus by the signal transduction pathway.

The first circuit functions as follows. Expression of the gene *GATA1* results in the appearance of the inactive transcription factor precursor GATA1inact in the cytoplasm (figure 7.5a). This precursor enters the cell nucleus and acts there in a phosphorylated form (GATA1-p) with its binding site in the promoter of the *GATA1* gene. The presence of GATA1 binding sites in the promoter of its own gene results in a rapid self-enhancement of its transcription (figure 7.5b) according to the positive feedback mechanism (Tsai et al., 1991). This positive feedback is very efficient and rapid because no other mediator genes are involved in its function.

The second positive feedback circuit, enhancing the *GATA1* gene transcription, functions in the following way. The GATA1 binding site is present in the promoter of the erythropoietin receptor (EPOR) gene (Chin et al., 1995). Transcription activation of this gene increases the number of erythropoietin receptor molecules on the cell membrane and, correspondingly, the intensity of signal transduction from erythropoietin via its receptor to the gene *GATA1*. Thus, this positive feedback circuit is closed, providing for additional self-enhancement of the *GATA1* gene transcription.

When this positive feedback brings the level of transcription factor GATA1-p to a threshold value, transcription of a cassette of genes is activated simultaneously. This cassette includes genes coding for (1) α- and β-subunits of hemoglobin (BG and AG in figure 7.7) (Gourdon et al., 1992; Jackson et al., 1995) and (2) heme biosynthesis enzymes (ALAS-E, ALAD, PBGD, URO-S, CPO, and FCH in figure 7.7) (Surinya et al., 1997; Kaya et al., 1994; Mignotte et al., 1989; Tanabe et al., 1997; Tugores et al., 1994). Transcription of the cassette of genes encoding cell surface antigens (GPD, GPB, and GPC) is also activated (Gregory et al., 1996; Colin et al., 1990; Iwamoto et al., 1996). Finally, transcription of the cassette of genes encoding transcription factors (HOXB2, TAL1, and EKLF in figure 7.7), particularly important for switching to differentiation, is activated (Lecointe et al., 1994; Vieille-Grosjean and Huber, 1995; Crossley et al., 1994). These factors provide for an additional stimulation of erythroid-specific genes. In this way a cascade of processes ensuring the terminal differentiation and maturation of the erythrocyte is triggered.

The cassette-like activation of genes, positive regulatory effects, and the key activators are observed in virtually all the gene networks described in the GeneNet database that trigger irreversible processes in the organism.

MATHEMATICAL SIMULATION OF GENE NETWORK FUNCTIONAL DYNAMICS

Before considering the main body of results, some introductory remarks on specifics of the objects simulated are necessary because they essentially determined our selection of the simulation tools.

First, even the simplest gene networks comprise dozens of physical components and interactions between them. Typically, gene networks are composed of hundreds of elements and more. Structural-functional organizations of many gene networks have been clarified in sufficient detail. Of the utmost importance is the fact that, despite their great diversity, the structural-functional organization of gene networks permits them to be directly and graphically represented by means of chemical kinetic description of elementary events constituting the gene network function.

Second, gene networks as actual objects exist simultaneously in a variety of different physical (material) forms due to such natural phenomena as gene polyallelism, genetic rearrangements, and performance of similar functions in different organisms. The more complex a gene network is, the wider the permissible diversity range of its variants. The number of variants is increased considerably by construction of artificial systems (such as expression vectors). In addition, gene networks themselves are elements of more complex biological systems.

Thus, the methods used for formalization should allow not only analysis of the initial gene networks and their minor modifications (mutations), but also calculation of the dynamics of virtually any genetic variants constructed from them. In other words, any model of a gene network should potentially contain models of many gene networks, including those drastically different from the initial one. A generalized chemical kinetic method, developed earlier and already applied to simulation of various gene networks (Bazhan et al., 1995; Belova et al., 1995; Likhoshvai et al., 2000), meets all these requirements. This method provides a precise and effective simulation of the gene network performance patterns because its computer realization adequately reflects the basic properties of biological systems.

BRIEF DESCRIPTION OF GENERALIZED CHEMICAL KINETIC SIMULATION METHOD

The generalized chemical kinetic simulation method is oriented to a formalized, primarily portrait, description of the performance patterns of arbitrary biological systems. Formalization is performed according to a block principle: a simulated system is divided into elementary subsystems in order to describe each subsystem individually. Elementary subsystems are described in terms of formal blocks. A formal block is uniquely characterized by an ordered list of formal dynamic variables $\bar{X}$, an ordered list of formal parameters $\bar{P}$, and the law of information transformation $\bar{F}$ (figure 7.8). The law of information transformation $\bar{F}$ can be uninterrupted, discrete, logical, or stochastic. The choice of elementary process description is determined by the nature of a biological process, the task to be solved, and the preference of a model's author. Therefore the models worked out within the generalized kinetic simulation method are generally hybrid (i.e., they can have uninterrupted, discrete, logical, and other blocks). The models that belong to the systems of common differential equations with a rational right-hand side are an important class of models for successful descriptions of gene network functioning. Such models appear when elementary blocks with differential equations expressing the immediate rate of concentration changes through immediate values of component concentration changes (mRNA, proteins, nonproteinaceous substrates, their complexes, etc.) in a simulated system are used as a law of information transformation.

N. A. Kolchanov et al.

1. Bimolecular reversible reaction : $x_1 + x_2 \underset{k_2}{\overset{k_1}{\Longleftrightarrow}} x_3$

$\overline{X} = (x_1, x_2, x_3), \quad \overline{P} = (k_1, k_2),$

$\overline{F} : \dfrac{dx_1}{dt} = k_2 \cdot x_3 - k_1 \cdot x_1 \cdot x_2, \quad \dfrac{dx_1}{dt} = \dfrac{dx_2}{dt} = -\dfrac{dx_3}{dt}.$

2. Monomolecular irreversible reaction : $x \xrightarrow{k} y$

$\overline{X} = (x, y), \quad \overline{P} = (k),$

$\overline{F} : \dfrac{dx}{dt} = -\dfrac{dy}{dt} = -k \cdot x.$

3. Michaelis – Menten scheme

$\overline{X} = (e, s, p), \quad \overline{P} = (k_o, k_m),$

$\overline{F} : \dfrac{ds}{dt} = -Z, \quad \dfrac{dp}{dt} = Z$

$Z = \dfrac{k_o es}{k_m + s}.$

4. Universal scheme

$\overline{X} = (x_1, \ldots, x_m, y_1, \ldots, y_n),$

$\overline{P} = (s_1, \ldots, s_l, a_1, \ldots, a_m, b_1, \ldots, b_n), m \geq 1, n \geq 0,$

$\overline{F} : \begin{cases} \dfrac{dx_j}{dt} = -a_j \cdot Z, \ j = 1, \ldots, m, \\[2mm] \dfrac{dy_l}{dt} = b_l \cdot Z, l = 1, \ldots, n \end{cases}$

where $Z = \dfrac{R(x_1, \ldots, x_m, s_1, \ldots, s_l)}{Q(x_1, \ldots, x_m, s_1, \ldots, s_l)},$

R, Q – polynomes from variable values **m**

Figure 7.8 Blocks of the Generalized Chemical Kinetic Simulation Method (GCKSM): $\overline{X}$, ordered list of formal variables; $\overline{P}$, ordered list of formal parameters; and $\overline{F}$, the law of transformation of information.

The chemical-kinetic approach is a methodological base for developing this class of models. The reason for its application is that biochemical reactions are the base of the processes controlled by gene networks. Mono- and bimolecular reactions are the base of the simplest biochemical reactions. The rate of changes in component concentration values in the reactions proceeding in a perfect mixture can be described with the systems of common and autonomous equations shown in figure 7.8,

General view of an equation setting the regularity of alteration of variable s rate

$dx_i/dt = F_i(x_1, ..., x_i, ..., x_n, k_1, ..., k_j, ..., k_m)$,

$i=1, ,n; x_i$ - model dynamic variables, k_j - constants.

1 reaction	2 reaction
$A + B <=> C$	$C => A + P$
$da/dt = \{-k_1 ab + k_2 c\}$,	$dc/dt = \{-k_3 c\}$,
$db/dt = \{-k_1 ab + k_2 c\}$,	$da/dt = -\{-k_3 c\}$,
$dc/dt = -\{-k_1 ab + k_2 c\}$.	$dp/dt = -\{-k_3 c\}$.

Model

$x_1 = a$: $da/dt = \{-k_1 ab + k_2 c\} - \{-k_3 c\} = F_1$,

$x_2 = b$: $db/dt = \{-k_1 ab + k_2 c\} \qquad\qquad = F_2$,

$x_3 = c$: $dc/dt = -\{-k_1 ab + k_2 c\} + \{-k_3 c\} = F_3$,

$x_4 = p$: $dp/dt = \qquad\qquad\qquad -\{-k_3 c\} = F_4$.

Figure 7.9 Rule of summing immediate velocities in elementary processes.

nos 1 and 2. It is not complicated to formulate a general system of equations even when several mono- and bimolecular biochemical reactions proceed in a medium simultaneously (figure 7.9).

It is necessary to use the rule of summing immediate velocities of the simplest reactions: the product of changes in immediate concentration velocities of the agent involved in several reactions is the sum of immediate velocity changes of the given agent in these reactions. As can be seen, the equation systems are quite simple when described only with mono- and bimolecular reactions.

Velocities of concentration changes are formulated by means of bilinear expressions from the same concentrations. In principle, two reaction types (bi- and monomolecular) are quite sufficient to describe arbitrary gene networks. However, such models will have a great many variables and parameters. In addition it is necessary to conduct biosystem decomposition on mono- and bimolecular reactions, although it is not always justified due to the lack of knowledge about concrete mechanisms of a particular gene network-controlled process. The natural way out of the deadlock is to consider more complicated processes as elementary and describe the laws of their proceeding in the form of autonomous equation systems with rational right-hand sides. The

N. A. Kolchanov et al.

Michaelis-Menten equation, frequently used in approximate descriptions of enzyme synthesis, is one of the most famous examples (figure 7.8, no. 3). More complicated reactions will be described with more complicated equations. The right-hand sides can be arbitrarily rational in a general case (figure 7.8, no. 4).

Since the rule of summing up immediate velocities does not depend on the internal complexity of elementary processes, it does not make the procedure of formulating a complete equation system more complicated. The total of agent concentration velocity changes is still the sum of the velocity changes in all elementary processes. It allows us to formulate a common approach to the description of regularities in the functioning of biological systems, particularly gene networks. The simulated system splits into simpler parts that will be considered elementary. Further on, elementary processes will be the building blocks used to develop the models not only of initial systems but also of their various modifications.

A great number of other models can be developed out of elementary processes in general, since any integration of an elementary process is a potential model. In particular it is possible to construct the models of practically arbitrary genetic variants of the initial network. There is no need to construct all the models at once; each particular model is constructed as required. Our method in its most general realization (Likhoshvai et al., 2000) implies the possibility of simulation not only of trans-interactions but also of cis-interactions. The latter should be taken into account in cases where patterns of gene network performance depend not only on the functions, but also on relative location, of the genes involved.

MATHEMATICAL MODEL OF REGULATION OF CHOLESTEROL BIOSYNTHESIS IN THE CELL

Cholesterol, an amphiphilic lipid, is an essential structural component of the cell membrane and the outer layer of the lipoproteins of blood plasma. Simultaneously it is a precursor of corticosteroids, sex hormones, bile acids, and vitamin D. Cholesterol is synthesized from acetyl-CoA, and its major fraction in the blood plasma is in the low-density lipoproteins (LDL). Cholesterol is removed from tissues with involvement of high-density lipoproteins (HDL) to be transported to the liver and transformed there into bile acids. In pathology, cholesterol is a factor causing atherosclerosis of vital cerebral arteries, heart muscle, and other organs. A high ratio of LDL cholesterol to HDL cholesterol in the plasma is observed in coronary atherosclerosis. This reveals the great biomedical and applied importance of studying cholesterol turnover in the organism.

The gene network regulating intracellular cholesterol biosynthesis has now been studied in sufficient detail. Data on its performance patterns are accumulated in the GeneNet database (`http://wwwmgs.bionet.nsc.ru/systems/mgl/genenet/`). Acetyl-CoA is the source of all the carbon atoms of the cholesterol molecule. The cholesterol biosynthesis pathway has numerous stages and is controlled by a variety of enzymes, including HMG-CoA reductase, farnesyl diphosphate synthetase, and squalene synthetase. Syntheses of these enzymes are activated by SREBP. The activity of SREBP depends, in turn, on the intracellular cholesterol concentration in a negative feedback mode: the lower the concentration of metabolically active cholesterol in the cell, the higher the SREBP activity.

We have developed a model of functional dynamics of this gene network. It describes all the stages of cholesterol biosynthesis shown in figure 7.4a as edges with adjacent nodes. In addition, the model describes the mechanisms underlying the interchange of intracellular and blood cholesterol. Negative feedbacks whereby cholesterol controls its own synthesis and the synthesis of LDL receptors at the transcription level (Wang et al., 1994) are also considered. The model fragment that consists of three equations describing the cycle of molecules acetyl-CoA, acetoacetyl-CoA, and HMG-CoA—three cholesterol precursors—is presented in figure 7.10. The equations are based on considering six processes in which they are agents. Right-hand equation members corresponding to one process are in braces and accompanied by indexes, displayed as subscripts, following the right brace: 1, entry of acetyl-CoA into the medium of the gene network functioning; 2, tiolase-catalyzed synthesis of acetoacetyl-CoA from two molecules of acetyl-CoA; 3 and 4, withdrawal of acetyl-CoA and acetoacetyl-CoA from the medium; 5, HMG-CoA synthase (HMGCS)-catalyzed synthesis of HMG-CoA from acetoacetyl-CoA and acetyl-CoA; 6, HMG reductase (HMGR)-catalyzed synthesis of mevalonic acid from HMG-CoA.

In totally, the model comprises 65 elementary processes. The model of performance dynamics of this gene network described in the GeneNet database contains 40 products (dynamic variables) and 93 constants. Values of a number of constants were assessed using the

$$d[Acetyl - CoA]/dt = \{k_{s,1}\}_1 + \{-k_{o,5}[HMG - CoA\ synthase][Acetyl - CoA][Acetoacetyl - CoA]/$$
$$(K_{m,5} + [HMG - CoA\ synthase][Acetyl - CoA] + [Acetyl - CoA][Acetoacetyl - CoA] +$$
$$[HMG - CoA\ synthase][Acetoacetyl - CoA])\}_5 + \{-k_{d,3}[Acetyl - CoA]\}_3$$

$$d[Acetoacetyl - CoA]/dt = \{k_{o,2}[Tiolase][Acetyl - CoA]^2 /(K_{m,2} + 2 \cdot [Tiolase][Acetyl - CoA] +$$
$$[Acetyl - CoA]^2)\}_2 + \{-k_{o,5}[HMG - CoA\ synthase][Acetyl - CoA][Acetoacetyl - CoA]/(K_{m,5} +$$
$$[HMG - CoA\ synthase][Acetyl - CoA] + [Acetyl - CoA][Acetoacetyl - CoA] +$$
$$[HMG - CoA\ synthase][Acetoacetyl - CoA])\}_5 + \{-k_{d,4}[Acetoacetyl - CoA]\}_4$$

$$d[HMG - CoA]/dt = \{k_{o,5}[HMG - CoA\ synthase][Acetyl - CoA][Acetoacetyl - CoA]/$$
$$(K_{m,5} + [HMG - CoA\ synthase][Acetyl - CoA] + [Acetyl - CoA][Acetoacetyl - CoA]$$
$$+ [HMG - CoA\ synthase][Acetoacetyl - CoA])\}_5 + \{-k_{o,6}[HMG\ reductase][HMG - CoA]/(K_{m,6} +$$
$$[HMG\ reductase] + [HMG - CoA])\}_6$$
$$\cdots\cdots$$

Figure 7.10 GeneNet model fragment controlling cholesterol biosynthesis.

relevant published data. The rest of the parameters were determined through numerical experiments using quantitative and qualitative characteristics known in the literature as criteria of their adequacies (Ratushny, Ignatieva, et al., 2000). This model allows the equilibrium state of the biosystem to be calculated. The equilibrium persists while the environmental conditions remain constant. If they change (e.g., the content of LDL particles in blood plasma increases twofold), the system equilibrium is disturbed (figure 7.11). Consequently the concentration of receptors bound to LDL increases (e) and that of receptors unbound to LDL decreases (d). Intracellular concentrations of free cholesterol (a) and its esters (c) increase. Unless a new intervention occurs, the negative feedbacks restore the initial state of the system: the initial cholesterol concentration in the cell is reestablished in approximately 3 hr, and the overall initial state of the system is restored in 10–15 hr.

MATHEMATICAL MODEL OF REGULATION OF ERYTHROCYTE MATURATION

Hematopoietic tissue belongs to the self-renewing systems of the organism that are operated through specific regulatory and self-regulatory mechanisms. Maintenance of a certain number of erythroid cells is

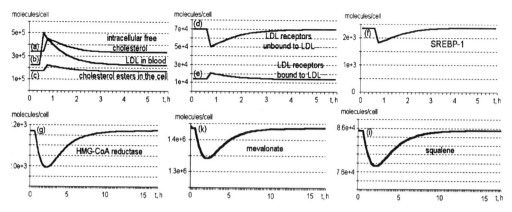

Figure 7.11 Kinetic changes in the main components of the system regulating cholesterol biosynthesis in the cell (a) in response to a simulated twofold increase (by 30 min) in LDL concentration in blood (b) x axis, time (h); y axis, molecules/cell. The main changes are the following: (d) the number of receptors unbound to LDL decreases; (e) the number of receptors bound to LDL increases; (a) intracellular concentration of free cholesterol and (c) its esters increase; (f) protease (SRP) binds free cholesterol, causing a decrease in SPEBP-1 concentration; (g) productions of the enzymes involved in the intracellular cholesterol synthesis (HMG-CoA reductase) stops; (k, i) production of LDL receptors and intermediate low-molecular-weight components (mevalonic acid, squalene) also stops. The system returns to the initial state until new input of exogenous cholesterol; concentration of free cholesterol in the cell restores approximately in 3 hours; complete restoration requires about 10–15 hours.

one of the necessary conditions for the organism to perform its vital functions. From this standpoint theoretical research into proliferation and differentiation of hematopoietic tissue cells is of both basic and applied biomedical importance.

The main stages of erythrocyte maturation are regulated by the gene network presented in figure 7.7. The hormone erythropoietin interacts with immature erythroid cells (erythroid stem progenitors of CFU-E type) and stimulates their proliferation, as well as syntheses of hemoglobin and the enzymes involved in heme biosynthesis, that is, maturation and differentiation of erythroid progenitors (Podkolodnaya et al., 2000). Low partial pressure of oxygen in venous blood (hypoxia) is another stimulator of erythropoietin synthesis.

Interacting with the cell receptor, erythropoietin activates the transcription factor GATA1, a key regulator of erythrocyte differentiation. GATA1 stimulates syntheses of α- and β-globins and the enzymes of heme biosynthesis. In addition it activates its own gene and the gene

N. A. Kolchanov et al.

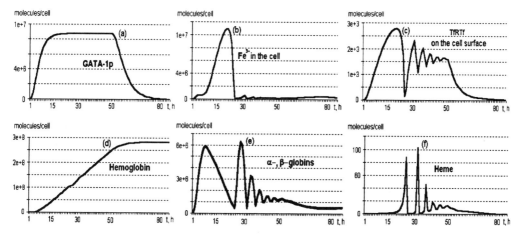

Figure 7.12 Dynamics of absolute concentrations of the main components of the erythroid cell differentiation system, calculated with x axis, time (h); y axis, molecules/cell. In (c), TfRTf-transferring receptors bound to transferring. Consecution of events: at beginning of erythropoietin influence on the cell, the precursor initiates the program of erythroid cell differentiation; 0–50 h is the total (transit) time of the function of erythroid cell differentiation in the gene network; at 50 h, the nucleus is lost; and 50–80 h, residual syntheses of the system's components in erythrocyte.

of the erythropoietin receptor (positive feedback). Heme, α-globin, and β-globin form hemoglobin, the major component of the mature erythrocyte.

This biosystem functions according to a pattern completely different from the system described in the previous section. This gene network is inactive until erythropoietin triggers it to start the irreversible process of erythrocyte maturation, in which positive feedbacks are predominant. The model of this gene network performance is described as a sequence of events occurring in the maturating cells of the erythroid lineage. The erythropoietin-responsive progenitor cell opens the lineage closed by the mature erythrocyte. The model comprises 119 elementary processes, 68 products, and 178 constants. Direct experimental data were used to specify a number of constants. The rest of the parameters were determined through numerical experiments using quantitative and qualitative characteristics known from the literature (Ratushny, Podkolodnaya, et al., 2000). The model developed predicts that production of several components of the system follows oscillatory dynamics (figure 7.12). This pattern results from interaction of positive and negative feedbacks, with the former regulation mode.

CONCLUSIONS

A limited set of elementary structures and events underlies the functions of gene networks; however, its elements combine to bring forth a tremendous diversity of existing gene networks and variants of their functions. Obligatory components of each gene network are (a) a group of genes expressed in a concerted manner (core of the gene network); (b) proteins encoded by these genes, which fulfill structural, transport, catalytic, regulatory, and other functions; (c) pathways transducing signals from cell membranes to cell nuclei, which provide for activation or suppression of gene transcription in response to stimuli external to the cell; (d) nonproteinaceous components that trigger the gene network function in response to external stimuli (hormones and other signal molecules); (e) various metabolites arising during the gene network functions (Kolpakov et al., 1998).

A characteristic feature of gene network organization is its capacity for self-regulation by means of closed regulatory circuits with negative and positive feedbacks (Kolchanov, 1997). These two types of regulatory circuits make it possible to maintain a definite functional state of the gene network or, on the contrary, enable its transition to another function mode under the effect of various factors, including environmental ones.

The gene network functions are regulated by hierarchically organized mechanisms whose cores are the key regulatory proteins coordinating the functions of the rest of the genes constituting a particular gene network. Groups of similar target sites in gene regulatory regions, capable of interacting with key regulatory proteins and thereby providing the cassette-type activation of large gene groups, form the molecular basis of the function of such regulatory circuits. This cassette-type activation of transcription of large gene groups is a characteristic feature of the gene networks described in the GeneNet database.

The hierarchical principle of the function of a gene network in a multicellular organism is the most important feature. Figure 7.13 qualitatively illustrates the hierarchy of mammalian gene networks. The lowest level of this hierarchy corresponds to gene networks controlling basal cell metabolism. Their functions depend on the stage of the cell cycle and can be suppressed or activated by regulatory effects coming from gene networks of higher levels. These effects can change the rates

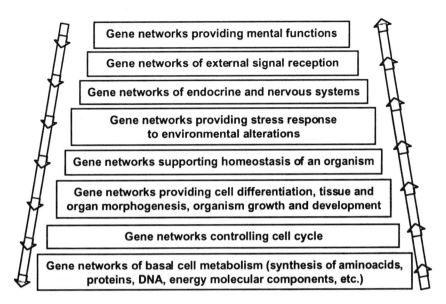

Figure 7.13 Global gene network of an organism (hierarchical integration).

and directions of both metabolic and cell division processes. The highest organization levels correspond to gene networks controlling external signal reception and mental functions.

Despite the simplicity of the classification described, it nevertheless allows the groups of gene networks with qualitatively similar functions to be detected, and their interaction and cosubordination to be described. In this global and hierarchically organized system of gene networks, the controlling signals are directed not only from higher levels to lower levels, but also in the opposite direction; and the signals may have both positive (activating) and negative (suppressing) effects.

The extreme complexity of gene networks requires application of mathematical simulation methods for investigating the networks' dynamics. The resulting models will allow the dynamics of both normal gene networks and those exhibiting pathologies, including those connected with gene mutations, to be studied. In addition to the possibilities such models present for basic studies of the gene network operation, they are very important for pharmacology and genetic engineering.

REFERENCES

Anan'ko, E. A., Bazhan, S. I., Belova, O. E., and Kel', A. E. (1997). Mechanisms of transcription of the interferon-induced genes: A description in the IIG-TRRD information system. *Molec. Biol. (Mosk)* 31: 592–605.

Bazhan, S. I., Likhoshvai, V. A., and Belova, O. E. (1995). Theoretical analysis of the regulation of interferon expression during priming and blocking. *J. Theor. Biol.* 175: 149–160.

Belova, O. E., Likhoshvai, V. A., Bazhan, S. I., and Kulichkov, V. A. (1995). Computer system for investigation and integrated description of molecular-genetic system regulation of interferon induction and action. *CABIOS* 11: 213–218.

Boyes, J., Byfield, P., Nakatani, Y., and Ogryzko, V. (1998). Regulation of activity of the transcription factor GATA-1 by acetylation. *Nature* 396: 594–598.

Briegel, K., Bartunek, P., Stengl, G., Lim, K. C., Beug, H., Engel, J. D., and Zenke, M. (1996). Regulation and function of transcription factor GATA-1 during red blood cell differentiation. *Development* 122: 3839–3850.

Chin, K., Oda, N., Shen, K., and Noguchi, C. T. (1995). Regulation of transcription of the human erythropoietin receptor gene by proteins binding to GATA-1 and Sp1 motifs. *Nucleic Acids Res.* 23: 3041–3049.

Colin, Y., Joulin, V., Le Van Kim, C., Romeo, P. H., and Cartron, J.-P. (1990). Characterization of a new erythroid/megakaryocyte-specific nuclear factor that binds the promoter of the housekeeping human glycophorin C gene. *J. Biol. Chem.* 265: 16729–16732.

Crossley, M., Tsang, A. P., Bieker, J., and Orkin, S. H. (1994). Regulation of the erythroid Kruppel-like factor (EKLF) gene promoter by the erythroid transcription factor GATA-1. *J. Biol. Chem.* 269: 15440–15444.

Dalyot, N., Fibach, E., Ronchi, A., Rachmilewitz, E. A., Ottolenghi, S., and Oppenheim, A. (1993). Erythropoietin triggers a burst of GATA-1 in normal human erythroid cells differentiating in tissue culture. *Nucleic Acids Res.* 21: 4031–4037.

Elliott, S., Lorenzini, T., Yanagihara, D., Chang, D., and Elliott, G. (1996). Activation of the erythropoietin (EPO) receptor by bivalent anti-EPO receptor antibodies. *J. Biol. Chem.* 271: 24691–24697.

Ericsson, J., Jackson, S. M., Lee, B. C., and Edwards, P. A. (1996). Sterol regulatory element binding protein binds to a cis element in the TI promoter of the farnesyl diphosphate synthase gene. *Proc. Natl. Acad. Sci. USA* 93: 945–950.

Gourdon, G., Morle, F., Roche, J., Tourneur, N., Joulain, V., and Godet, J. (1992). Identification of GATA-1 and NF-E2 binding sites in the flanking region of the human alpha-globin genes. *Acta Haematol.* 87: 136–144.

Gregory, R. C., Taxman, D. J., Seshasayee, D., Kensinger, M. H., Bieker, J. J., and Wojchowski, D. M. (1996). Functional interaction of GATA1 with erythroid Kruppel-like factor and Sp1 at defined erythroid promoters. *Blood* 87: 1793–1801.

Hofestadt, R., and Meineke, F. (1995). Interactive modelling and simulation of biochemical networks. *Comput. Biol. Med.* 25: 321–334.

Ignatieva, E. V., Merkulova, T. I., Vishnevskii, O. V., Kel', A. E. (1997). Transcriptional regulation of lipid metabolism genes: description in the TRDD database. *Molec. Biol. (Mosk)* 31: 575–591.

Iwamoto, S., Li, J., Sugimoto, N., Okuda, H., and Kajii, E. (1996). Characterization of the Duffy gene promoter: Evidence for tissue-specific abolishment of expression in Fy(a-b-) of black individuals. *Biochem. Biophys. Res. Comm.* 222: 852–859.

Jackson, C. E., O'Neill, D., and Bank, A. (1995). Nuclear factor binding sites in human beta globin IVS2. *J. Biol. Chem.* 270: 28448–28456.

Kauffman, S. A. (1969). Metabolic stability and epigenesis in randomly constructed genetic net. *J. Theor. Biol.* 22: 437–467.

Kaya, A. H., Plewinska, M., Wong, D. M., Desnick, R. J., and Wetmur, J. G. (1994). Human delta-aminolevulinate dehydratase (ALAD) gene: Structure and alternative splicing of the erythroid and housekeeping mRNAs. *Genomics* 19: 242–248.

Kolchanov, N. A. (1997). Regulation of the eukaryotic gene transcription: Databases and computer analysis. *Mol. Biol.* 31: 481–482.

Kolchanov, N. A., Podkolodnaya, O. A., Ananko, E. A., Ignatieva, E. V., Stepanenko, I. L., Kel-Margoulis, O. V., Kel', A. E., Merkulova, T. I., Goryachkovskaya, T. N., Busygina, T. V., Kolpakov, F. A., Podkolodny, N. L., Naumochkin, A. N., Korostishevskaya, I. M., Romashchenko, A. G., and Overton, G. C. (2000). Transcription Regulatory Regions Database (TRRD): Its status in 2000. *Nucleic Acids Res.* 28: 298–301.

Kolpakov, F. A., and Ananko, E. A. (1999). Interactive data input into the GeneNet database. *Bioinformatics* 15: 713–714.

Kolpakov, F. A., Ananko, E. A., Kolesov, G. B., and Kolchanov, N. A. (1998). GeneNet: A database for gene networks and its automated visualization. *Bioinformatics* 14: 529–537.

Lecointe, N., Bernard, O., Naert, K., Joulin, V., Larsen, C. J., Romeo, P. H., and Mathieu-Mahul, D. (1994). GATA- and SP1-binding sites are required for the full activity of the tissue-specific promoter of the tal-1 gene. *Oncogene* 9: 2623–2632.

Likhoshvai, V. A., Matushkin, Yu. G., Vatolin, Yu. N., and Bazhan, S. I. (2000). A generalized chemical kinetic method for simulating complex biological systems. A computer model of λ phage ontogenesis. *Computational Technol.* 5(2): 87–99.

Lloyd, D. B., and Thompson, J. F. (1995). Transcriptional modulators affect in vivo protein binding to the low density lipoprotein receptor and 3-hydroxy-3-methylglutaryl coenzyme A reductase promoters. *J. Biol. Chem.* 270: 25812–25818.

McAdams, H. H., and Arkin, A. (1997). Stochastic mechanism in gene expression. *Proc. Natl. Acad. Sci. USA* 94: 814–819.

Merkulova, T. I., Merkulov, V. M., and Mitina, R. L. (1997). Mechanisms of glucocorticoid regulation and regulatory regions of genes, controlled by glucocorticoids: description in the TRDD database. *Molec. Biol. (Mosk)* 31: 714–725.

Mignotte, V., Wall, L., deBoer, E., Grosveld, F., and Romeo, P. H. (1989). Two tissue-specific factors bind the erythroid promoter of the human porphobilinogen deaminase gene. *Nucleic Acids Res.* 17: 37–54.

Podkolodnaya, O. A., and Stepanenko, I. L. (1997). Mechanisms of transcriptional regulation of erythroid specific genes. *Molec. Biol. (Mosk)* 31: 562–574.

Podkolodnaya, O. A., Stepanenko, I. L., Ananko, E. A., and Vorobiev, D. G. (2000). Representation of information on erythroid gene expression regulation in the Geneéxpress system. In proceedings of the Second International conference on Bioinformatics or genome Regulation and Structure. Novosibirsk: ICG, pp. 34–36.

Ratner, V. A. (1966). *Genentical Controlling Systems*. Novosibirsk: Nauka.

Ratushny, A. V., Ignatieva, E. V., Matushkin, Yu. G., and Likhoshvai, V. A. (2000). Mathematical model of cholesterol biosynthesis regulation in the cell. In proceedings of the Second International Conference on Bioinformatics or Genome Regulation and Structure. Novosibirsk: ICG, pp. 199–202.

Ratushny, A. V., Podkolodnaya, O. A., Ananko, E. A., and Likhoshvai, V. A. (2000). Mathematical model of erythroid cell differentiation regulation. Proceeding of the second international conference on bioinformatics or genome regulation and structure. Novosibirsk: ICG, pp. 203–206.

Sanchez, L., van Helden, J., and Thieffry, D. (1997). Establishement of the dorso-ventral pattern during embryonic development of *Drosophila melanogaster*: A logical analysis. *J. Theor. Biol.* 189: 377–389.

Savageau, M. A. (1985). A theory of alternative designs for biochemical control systems. *Biomed. Biochim. Acta.* 44: 875–880.

Schweigert, S., Herde, P. V., and Sibbald, P. R. (1995). Issues in incorporation of semantic integrity in molecular biological object-oriented databases. *Comput. Appl. Biosci.* 11: 339–347.

Sugita, M. (1961). Functional analysis of chemical systems in vivo using a logical circuit equivalent. *J. Mol. Biol.* 1: 415–430.

Surinya, K. H., Cox, T. C., and May, B. K. (1997). Transcriptional regulation of the human erythroid 5-aminolevulinate synthase gene. Identification of promoter elements and role of regulatory proteins. *J. Biol. Chem.* 272: 26585–26594.

Tanabe, A., Furukawa, T., Ogawa, Y., Yamamoto, M., Hayashi, N., Tokunaga, R., and Taketani, S. (1997). Involvement of the transcriptional factor GATA-1 in regulation of expression of coproporphyrinogen oxidase in mouse erythroleukemia cells. *Biochem. Biophys. Res. Comm.* 233: 729–736.

Tchuraev, R. N. (1991). A new method for the analysis of the dynamics of the molecular genetic control systems. I. Description of the method of generalized threshold models. *J. Theor. Biol.* 151: 71–87.

Thieffry, D., and Thomas, R. (1995). Dynamical behaviour of biological regulatory networks—II. Immunity control in bacteriophage lambda. *Bull. Math. Biol.* 57: 277–297.

Thomas, R. (1973). Boolean formalization of genetic control circuits. *J. Theor. Biol.* 42: 563–585.

Tsai, S.-F., Strauss, E., and Orkin, S. H. (1991). Functional analysis and in vivo footprinting implicate the erythroid transcription factor GATA-1 as a positive regulator of its own promoter. *Genes Dev.* 5: 919–931.

Tugores, A., Magness, S. T., and Brenner, D. A. (1994). A single promoter directs both housekeeping and erythroid preferential expression of the human ferrochelatase gene. *J. Biol. Chem.* 269: 30789–30797.

Vieille-Grosjean, I., and Huber, P. (1995). Transcription factor GATA-1 regulates human HOXB2 gene expression in erythroid cells. *J. Biol. Chem.* 270: 4544–4550.

Wahde, M., and Hertz, J. (2000). Coarse-grained reverse engineering of genetic regulatory networks. *Biosystems* 55: 129–136.

Wang, X., Pai, J.-T., Wiedenfeld, E. A., Medina, J. C., Slaughter, C. A., Goldstein, J. L., and Brown, M. S. (1995). Purification of an interleukin-1 beta converting enzyme-related cysteine protease that cleaves sterol regulatory element-binding proteins between the leucine zipper and transmembrane domains. *J. Biol. Chem.* 270: 18044–18050.

Wang, X., Sato, R., Brown, M. S., Hua, X., and Goldstein, J. L. (1994). SREBP-1, a membrane-bound transcription factor released by sterol regulated proteolysis. *Cell* 77: 53–62.

Zhang, W., and Bieker, J. J. (1998). Acetylation and modulation of erythroid Kruppel-like factor (EKLF) activity by interaction with histone acetyltransferases. *Proc. Natl. Acad. Sci. USA* 95: 9855–9860.

SUGGESTED READINGS

Kanehisa, M. (2000). *Post-Genome Informatics*. Oxford: Institute for Chemical Research, Kyoto University, and Oxford University Press.

Kohn, K. W. (1999). Molecular interaction map of the mammalian cell cycle control and DNA repair systems. *Mol. Biol. Cell* 10: 2703–2734.

Savageau, M. A. (1991). Biochemical systems theory: Operational differences among variant representations and their significance. *J. Theor. Biol.* 151: 509–530.

Smolen, P., Baxter, D. A., and Byrne, J. H. (2000). Modeling transcriptional control in gene networks—methods, recent results, and future directions. *Bull. Math. Biol.* 62: 247–292.

Thomas, R., Thieffry, D., and Kaufman, M. (1995). Dynamical behaviour of biological regulatory networks—I. Biological role of feedback loops and practical use of the concept of the loop-characteristic state. *Bull. Math. Biol.* 57: 247–276.

URLs FOR RELEVANT SITES

Cell Signaling Networks Database (CSNDB) is a data- and knowledge base for signaling pathways of human cells. It compiles the information on biological molecules, sequences, structures, functions, and reactions that transfer the cellular signals. Signaling pathways are compiled as binary relationships of biomolecules and are represented by automatically drawn graphs. `http://geo.nihs.go.jp/csndb/`

EcoCyc, the Encyclopedia of *E. coli* Genes and Metabolism, is a bioinformatics database that describes the genome and the biochemical machinery of *E. coli*. The long-term goal of the project is to describe the molecular catalog of the *E. coli* cell, as well as the functions of each of its molecular parts, in order to facilitate a system-level understanding of *E. coli*. `http://ecocyc.pangeasystems.com/ecocyc/`

ERGO is a curated database of genomic DNA with connected similarities, functions, pathways, functional models, clusters, and more. Users may annotate and comment on genes and pathways, but currently cannot edit sequences. `http://igweb.integratedgenomics.com/IGwit/`

GeNet contains the information on functional organization of regulatory gene networks acting at embryogenesis. The regulatory genes play a crucial role in embryogenesis, controlling both activity of downstream regulatory genes (cross-regulation) and their own activity (autoregulation). These are the autoregulatory and cross-regulatory functional links, which unite the regulatory genes in gene networks. `http://www.csa.ru/Inst/gorb_dep/inbios/genet/genet.htm`

Kyoto Encyclopedia of Genes and Genomes (KEGG) is an effort to computerize current knowledge of molecular and cellular biology in terms of the information pathways that consist of interacting molecules or genes, and to provide links from the gene catalogs produced by genome sequencing projects. `http://www.genome.ad.jp/kegg/`

The Ligand Chemical Database for Enzyme Reactions (LIGAND) is designed to provide the linkage between chemical and biological aspects of life in the light of enzymatic reactions. The database consists of three sections: ENZYME, COMPOUND, and the REACTION. The PATHWAY data item contains the link information to the KEGG (see above) metabolic pathway database: the pathway map accession number, followed by the description. `http://www.genome.ad.jp/dbget/ligand.html`

MetaCyc, the metabolic Encyclopedia, is a meta-metabolic database because it contains pathways from a variety of different organisms. It describes metabolic pathways, reactions, enzymes, and substrate compounds. The MetaCyc data were gathered from a variety of literature and on-line sources, and contain citations to the source of each pathway. MetaCyc employs the same database schema as EcoCyc and provides the same rich annotation for many of the pathways, based on the biomedical literature. Unlike EcoCyc, MetaCyc does not provide genomic data. It is also based on the same retrieval and visualization software as EcoCyc, the Pathway Tools. `http://ecocyc.pangeasystems.com/ecocyc/metacyc.html`

The Signaling PAthway Database (SPAD) is an integrated database for genetic information and signal transduction systems. SPAD is divided into four categories based on extracellular signal molecules—(growth factor, cytokine, and hormone) and stress—that initiate the intracellular signaling pathway. SPAD is compiled in order to describe infor-

mation on interaction between protein and protein, and protein and DNA, as well as information on sequences of DNA and proteins.
http://www.grt.kyushu-u.ac.jp/spad/

The University of Minnesota Biocatalysis/Biodegradation Database contains information about microbial biocatalytic reactions and biodegradation pathways primarily for xenobiotic, chemical compounds. http://umbbd.ahc.umn.edu/

WIT is a Web-based system to support the curation of function assignments made to genes and the development of metabolic models. http://wit.mcs.anl.gov/WIT2

Regulation of Cellular States in Mammalian Cells from a Genomewide View

Sui Huang

INTRODUCTION

The development from a single fertilized oocyte to a mature multicellular organism, as well as the maintenance of the complex adult tissue architecture consisting of cells of distinct characteristics, requires a tight temporal and spatial regulation of the "fate" of each cell, such as division, differentiation, and death. That the fate of an individual cell is subordinated to the needs of the collective is most impressively manifest in the finding that the majority of cells in the mature tissue are not in the proliferation mode despite the abundance of nutrients and the optimal physicochemical conditions. Only in regenerative tissues such as skin, intestinal epithelium, and bone marrow can a subset of cells switch to the proliferation mode. Disturbance of the balance between cell proliferation, differentiation, and death can lead to tumor formation.

Signal Transduction in a Multicellular Organism as Information Processing

What tells a cell in a multicellular organism when to divide, differentiate, or die? The signals involved in higher multicellular organisms are fundamentally different from those in single-cell organisms, such as bacteria and yeast, which have traditionally served as model systems for the study of molecular control circuits. In the microorganisms, regulation is geared toward optimizing material and energy utilization and stimulation of cell division, thus propagating the organism whenever environmental conditions are favorable. There is no subordination to the needs of the whole organism. A well-studied control mechanism is the lactose operon system in *E. coli*, which allows the bacterial cell to switch its metabolism from glucose to lactose as the source of free energy (for a review see Reznikoff, 1992).

In contrast to free-living microorganisms, cells of multicellular organisms in situ do not face the problem of having to adapt to fluctuating environmental conditions and varying types of energy sources. A liver cell, for instance, always sits in an environment in which temperature, pH, and concentration of nutrients (e.g., glucose) stay within narrow ranges. Here regulation serves tissue homeostasis (i.e., the maintenance of the intricate tissue microarchitecture that is the basis of organ function). Therefore, the regulated variable is the individual cell's decision between cell proliferation, differentiation, and death. While a lactose molecule acts in *E. coli* both as nutrient and as signal to activate the appropriate genetic program, messenger molecules for mammalian cells are complex peptides (e.g., "growth factors") that serve solely to convey a regulatory signal (e.g., to stimulate the cell to differentiate or divide).

The cell is not interested in the free energy present in the peptide bonds and amino acids of the growth factor. For a mammalian cell in situ, free energy is not rate-limiting, and the task of acquiring energy is delegated to a hierarchically higher level, the organism. Therefore, we should speak of a *regulatory network* instead of a metabolic network when referring to the network of genes and enzymes that governs cell fate in multicellular organisms. As with other complex systems, such as computers (as opposed to combustion engines) or modern societies (as opposed to agricultural societies), with increasing complexity the currency on which the system operates shifts from energy to *information* when the former is in abundant supply. Therefore, in a first abstraction, the web of information exchange can be studied as a separate network, although it is ultimately connected to the energy-supplying system.

Cell Fates as Cell States

In a simplified picture that will suffice for our discussion, we consider only cell proliferation, differentiation, and death as the cell fates. From a formal viewpoint we shall treat these cell fates as *cell states* rather than processes: "Proliferation" (or growth) is the state of biochemical competence for repeated cell divisions (an oscillatory state). "Differentiation" is the state in which the cell maintains its specialized, type-characteristic features and engages in tissue-specific tasks. A cell can have multiple differentiation states (e.g., the resting and the activated states of a lymphocyte), but for simplicity we will deal with just one. Although differentiated cells typically withdraw from the cell division cycle, and thus are in a quiescent state, we sometimes use the term "quiescence" separately to describe the state when considering exit from the proliferation mode independent of a differentiation process (e.g., dormancy, or arrest at a cell-cycle checkpoint for DNA repair). Finally, "programmed cell death" (or apoptosis) describes the state of commitment to apoptosis with the ensuing processes being merely a stepwise execution of cell death. The term "state" captures the essential features of physiological cell fates because they are stable, qualitatively distinct, and mutually exclusive, and can undergo transitions from one to another under the influence of regulatory inputs.

Interestingly, unlike bacteria, for mammalian cells in culture, supplying basic nutrients such as amino acids, fatty acids, and glucose, and keeping temperature, osmolarity, and pH within the physiological range, is not sufficient for the maintenance of viability and continued cell division. In addition, the culture medium needs to be supplemented with 2–20% serum to keep the cells alive and proliferating. We now know that this requirement for serum is essentially due to the presence of *growth factors* in the serum. The need for growth factors to keep cells alive and proliferating again illustrates the fact that regulation in mammalian cells takes place at the level of information processing and not of energy utilization.

Taken together, for the study of regulatory networks in cells of higher organisms we can in a first approximation regard cells as open (nonequilibrium) systems with a continuous supply of free energy, and thus focus on the regulatory events as information processing that leads to the decision either to maintain or to switch a cell state.

Traditional Paradigm of Cell Fate Regulation: A Pathway-centered View

The advent of molecular biology techniques has stimulated unprecedented progress in the biochemical analysis of cell fate regulation that started with the molecular cloning of growth factors and their receptors, and led to today's bewildering, complex picture of the molecular machinery that mediates the cellular response to these growth factors. The general approach to elucidating the molecular mechanism by which a growth factor determines cell fate consisted of breaking down the intracellular biochemical processes into individual *signal transduction pathways* that link the activation of a cell surface receptor by the growth factor to the induction of gene expression in the nucleus.

The basic idea of this pathway concept is that the instruction for a cell is encoded in the molecular structure of the extracellular ligand and its specific recognition by its cognate receptor. This key-lock model also applies to the ensuing intracellular events (Pawson and Scott, 1997). The external instruction is passed down to the nucleus by a chain of molecular recognition and (in)activation events that constitutes a signal transduction pathway: a chain of events connects cause and effect. The current state of our knowledge is typically summarized in the kind of "arrow-arrow" diagram in figure 8.1, where the flow of information starts on the top, with cell surface receptor activation, and ends at the bottom, with gene induction. Branching and cross connections between the nominal pathways are increasingly built-in in such diagrams; feedback loops are only rarely considered, however, so the pathways remain mostly *linear*. Moreover, their final targets are usually vaguely designated "gene induction." How these genes then collectively contribute to affecting the observed dynamic of cell fate remains unanswered by such schematic representations.

Analysis of entire genome sequences and large-scale gene expression studies in lower metazoan and mammalian cells indicates that more than 10% of the genome and up to 30% of the transcriptome consist of genes that encode proteins involved in signal transduction (Chervitz et al., 1998; Phillips et al., 2000; Venter et al., 2001). This high number of signaling genes underscores the importance of information processing. At the threshold to the postgenomic era of biology, we have now to start to ask: Can cloning and characterizing all these signaling proteins, and categorizing them into receptor-to-gene pathways, ultimately lead to an encompassing, conceptual understanding of the principles of cell fate regulation? More concretely, the central question is: How can we map the current signaling pathway schemes onto the real dynamics of cell behavior that consists of relatively few, distinct, mutually exclusive cell fates, such as growth, differentiation, and death?

These questions are of central significance because the present ad hoc "arrow-arrow" cartoons are challenged by accumulating observations

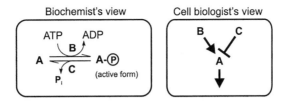

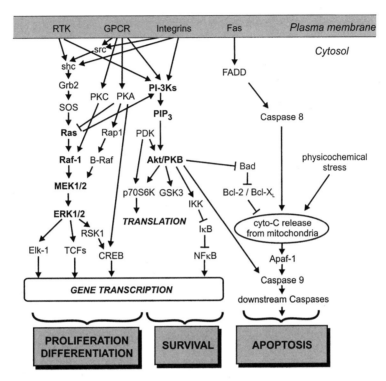

Figure 8.1 The current signal transduction paradigms in cell biology are based on "arrow-arrow" diagrams. The insets on top show how the biochemist's formal representation of an underlying reaction (left) is "translated" into the cell biologist's shorthand notation (right). Molecule A is activated by phosphorylation, catalyzed by the kinase B, and is inactivated by dephosphorylation, catalyzed by the phosphatase C. Inhibition might also be achieved by a molecule that binds to B and prevents its catalytic action on A.

The bottom panel illustrates how in current cell and molecular biology the established signaling pathways thought to govern the various cell fates are typically represented, based on the above shorthand notation. Only the most salient, established molecules and connections are shown. The canonical mitogenic pathway is depicted in boldface, consisting of the sequence Ras-Raf1-MEK1/2-ERK1/2, the classical MAP kinase pathway. ERK belongs to the MAP (mitogen-activated protein) kinases, which form a family of highly conserved mediators of diverse extracellular signals to the nucleus. The more recently described, central pathway for survival involves the PI3K-Akt sequence, also shown in boldface. Note the cross talk between the classical pathways of Ras, PI3K, PKA and PKC. All reactant labels are generic names of the proteins, except for the following abbreviations: RTK, receptor tyrosine kinase, GPCR, G-protein coupled receptor, PIP3, phosphatidyl-inositol-3, 4, 5-triphosphate, cyto-C, cytochrome c.

that cannot be accommodated by such complicated collections of individual pathways. These findings indicate that there are regulatory principles that can be conceived only in the broader picture of global cellular regulation. One such principle is what can be summarized as "distributed information" in signaling, and is manifest in the intense cross talk between the nominal pathways, in the pleiotropic effects of signaling molecules, and in the fact that regulatory cellular proteins often have multiple, opposing effects on cell fate (see below). These phenomena, albeit found ubiquitously, are presently discussed as complications or even paradoxes within our mental framework of linear, point-to-point pathways.

In this chapter we briefly review the limitations of the current pathway paradigm and demonstrate how the concept of Boolean networks can serve as a simple modeling language to formalize the emergence of cell fates as collective behavior of signaling molecules and to predict the functional consequences. Thus, we will introduce the idea of Boolean networks and their fundamental properties as far as it pertains to cell fate regulation, but will then put emphasis on building the bridge between the dynamics of the modeled network and the observed behavior of the real cell.

DISTRIBUTED INFORMATION PROCESSING IN CELL FATE REGULATION

The habit of assigning "functions" to biological molecules and pathways, as exemplified by terms like "fibroblast growth factor" and "mitogenic pathway," which has culminated in the creation of "functional genomics," needs to be contrasted with accumulating evidence that the information conveyed by the signal transduction machinery cannot always be localized to a particular pathway, let alone a molecule, but is distributed among numerous pathways that collectively form a genome–wide regulatory network. This has important implications for functional annotation of protein databases (as discussed in chapter 2). While assigning biochemical functions based on sequence analysis is relatively straightforward, many higher-order biological functions, such as "growth promotion," are often not intrinsic properties of individual proteins but depend on the cellular and tissue context (as discussed below). Such "conditional functions" might require an additional layer in protein function annotations, but certainly suggest that manual annotation based on experimental data will remain an important aspect of database annotation.

Cross Talk Between Pathways: Convergence and Pleiotropy

As the collection of known biochemical pathways has grown, it has become clear that an intense cross talk takes place between the historically defined pathways. New physical interactions between known genes or proteins, often from apparently unrelated "pathway systems," are published almost weekly. With regard to the scheme of interactions between nominal pathways, two forms of cross communication can be distinguished: convergence and pleiotropy (figure 8.2).

Convergence describes the finding that different pathways impinge on a same key target molecule. For instance, mitogenic signals from the cell's environment converge on the activation of the Ras-Raf-ERK pathway (figure 8.1), which plays a critical role in the activation of genes that stimulate entry into the cell cycle and drive progression of the G1 phase into the S phase (DNA synthesis phase) (Lavoie et al., 1996; Taylor and Shalloway, 1996). Similarly, signals from a variety of stimuli necessary for cell survival (prevention of cell death) utilize the PI3K-akt/PKB pathway (Franke et al., 1997; Downward, 1998; Krasilnikov, 2000). One important functional consequence of convergence is *redundancy*, in that the activation of a key molecule, such as the mitogen-activated protein kinase (MAPK) ERK, can be achieved by various signals and that has been viewed as molecular basis for the robustness of signal transmission. However, the fact that multiple non–identical input signals impinge on the very same mediator has raised the question of how signal specificity is maintained (Chao, 1992; Brunet and Pouysségur, 1997; Tan and Kim, 1999).

Pleiotropy, on the other hand, refers to the finding that activation of one molecule results in the fanning out of the biochemical signal, resulting in the induction of a large array of genes. For instance, monitoring of >6000 genes with DNA microarrays showed that activated receptors for the growth factors FGF and PDGF induce the expression of at least 60 genes (Fambrough et al., 1999) and that the Myc transcription factor activates at least 27 genes (Coller et al., 2000). The functional meaning of pleiotropy is not immediately plausible, since it also tends to lessen signal specificity through overlap of the responses of different input signals. However, because of their prevalence, con-

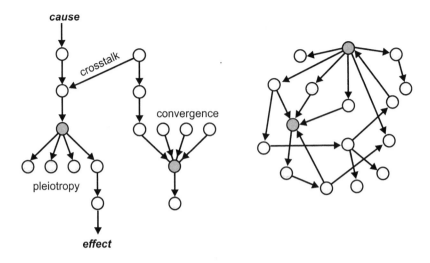

Traditional pathway model

Network model

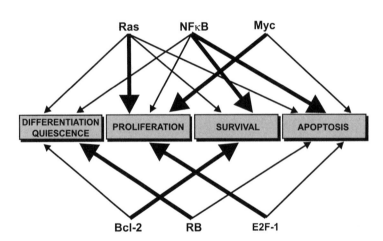

Figure 8.2 Distributed information processing in signal transduction is embodied in the various structures of interaction shown in the upper left: cross talk between the linear pathways, convergence, and pleiotropy. Although these features are appreciated by the traditional pathway paradigm, a nominal pathway consists mainly of a chain of events connecting cause to effect. A network model (upper right) of signal transduction is proposed here in which there is no single cause, and the effect is an emergent property of the networked interactions. Circles indicate molecules, and arrows denote catalytic activity upon another molecule, in the sense of figure 8.1. The bottom panel shows examples of signaling proteins that exhibit disparate effects on cell fate, another manifestation of distributed information processing. The thick arrows denote the original function first assigned to the protein (for details see Huang and Ingber, 2000a).

vergence and pleiotropy appear to represent basic design principles of signal transduction in cell fate regulation, and will be discussed in the context of regulatory networks later in this chapter.

Regulatory Proteins with Multiple Disparate Functions

The continuing analysis of key signaling molecules has revealed that many of them exert multiple, often disparate effects on cell fate. For instance, the protein Ras, originally discovered on the basis of its transforming (tumor-inducing) effect (Barbacid et al., 1987), not only promotes growth but also induces apoptosis and quiescence—an apparent paradox (figure 8.2). Ras promotes growth via the MAP kinase ERK, but, in the absence of the proteins NF·B and PI3K, its activation induces programmed cell death (Mayo et al., 1997; Kauffmann-Zeh et al., 1997; Gire et al., 2000).

The transforming activity of Ras requires cooperation with other immortalizing alterations, such as activation of the protein Myc (Weinberg, 1989). The notion of Ras being a growth-promoting protein stems from the use of *immortalized cell lines* during its initial characterization. In *primary cells* Ras fails to stimulate growth, but instead induces accumulation of the cell cycle-inhibiting proteins p16 and p53 and puts cells into quiescence (Serrano et al., 1997). Another known example is Myc, also first characterized as an oncogene. Myc protein promotes cell growth in the presence of serum or growth factors but triggers p53-dependent apoptosis in their absence (Evan and Littlewood, 1998). Figure 8.2 shows some other selected examples of proteins with disparate effects on cell fate (for details see Huang and Ingber, 2000a).

Multiple-Target or Nonspecific Perturbation

A third line of observation advocating a global (genomewide) approach to cell fate regulation is the readiness with which a specific cell fate can be induced, using an experimental perturbation that affects multiple proteins across several nominal pathways. This contrasts with the current paradigm of simple pathways, which is derived from the idea of *instructive* regulation: the extracellular factor acts as a messenger that carries in its unique three-dimensional structure the specific instruction to be delivered to the cell by docking to the complementary receptor. However, many small-molecule chemicals that do not act on one specific receptor, but on multiple targets, are equally able to cause a distinct cell state switch—not just by triggering apoptosis, which one could attribute to "nonspecific cytotoxicity," but also by inducing full differentiation with all the intricate cellular features.

Small organic molecules such as dimethylsulfoxide (DMSO), ethanol, beta-mercaptoethanol, benzene, and genistein (a general tyrosine kinase inhibitor) have been reported, for a large number of cell culture systems, to switch cells from growth to differentiation (Yu and Quinn, 1994; Kulyk and Hoffman, 1996; Woodbury et al., 2000; Kalf and O'Connor, 1993; Constantinou and Huberman, 1995). It appears that in all these examples a simultaneous perturbation of multiple targets in different pathways results in the cell's channeling the biochemical response in such a way that a well-orchestrated pattern of biochemical activity arises that drives the cell toward a distinct cell fate.

Even external signals as "nonspecific" as physical inputs, such as externally enforced change of cell shape, can regulate the choice of cell fate, switching the cell between quiescence, growth, differentiation, and apoptosis (Mooney et al., 1992; Chen et al., 1997; Huang and Ingber, 1999). The view held by most molecular biologists—that the cell is instructed on how to behave by the message encoded in the structure of regulatory molecules—cannot explain how these physical influences, devoid of molecular specificity, can trigger a distinct cell fate.

Toward an Integrative View Based on Genomewide Networks

From the above examples it is clear that dissecting cell fate regulation into individual pathways is problematic because the functional effects of individual regulatory proteins are not intrinsic properties of the proteins themselves, but instead depend on the cellular context (i.e., the presence and activity of other cellular components). Hence, information is distributed. It is the cell that integrates it into a distinct cell fate.

In recent attempts to bring some order to the richly intertwined web of signal transduction pathways, the concept of modularity has been proposed (Hartwell et al., 2000). Moreover, it has been suggested that protein modules (multiprotein complexes) with multiple inputs and outputs act as information processors (Bray, 1995). These suggestions preserve the convenient and intuitively strong notion of individual, separable pathways to which "functions" can be attributed without embracing a global view. Whether or not there is a physical basis for an a priori assumption of functional modules beyond the subjective, operational definition by the investigator remains open.

One could envisage that the generally sparse interconnections (see below) and the heterogeneity of cross-talk density throughout the genomic regulatory network might justify the idea of modularity. However, given the prevalence of unaccounted-for cross communication between "pathways," we propose here an approach to cellular regulation without an a priori notion of "functions" and modules, but with a neutral assumption of a *global network* with distributed information in which modularity emerges as a secondary entity—depending on the "function" we are looking at. Further, we will show in this chapter how the network approach can help resolve the questions on cross talk, disparate effects of proteins, nonspecific stimuli, and the preservation, despite convergence and pleiotropy, of signal specificity.

BOOLEAN NETWORKS AS A MODEL FOR CELLULAR REGULATORY NETWORKS

A Qualitative Modeling Language

With the recent explosion of biological data on cell regulation, it has become clear even to experimentalists that the counterintuitive, often paradoxical features and the network nature of cellular regulation exceed the explanatory capacity of the current ad hoc, schematic models used in cell and molecular biology. Mathematical models, or at least more formal methods, will be necessary to support our efforts to make sense of the continuing spate of experimental data. Traditional mathematical modeling approaches typically aim at recapitulating the precise kinetics of rather small regulatory circuitries using sets of differential equations. Other approaches pursue the simulation of the living organism (or parts of it) as realistically as possible at various scales of description, down to using stochastic methods to track individual molecules. All these modeling approaches have provided valuable insight into many nonintuitive behaviors that are typical for nonlinear systems, such as oscillations, multistability, and hysteresis. A motivation behind these approaches is that detailed simulations may one day serve as prediction tools that could replace experimentation (McAdams and Arkin, 1996; Tyson, 1999; Weng et al., 1999; Endy and Brent, 2001).

However, these approaches do not aim at understanding basic design principles of regulation, and the problems discussed in the previous section have not been addressed. To study the regulation of cell fate as a collective behavior mediated by distributed information processing, we have to deal with a massively parallel, genome wide system of interacting parts and learn to describe how they give rise to the distinct, emergent behavior apparent at a higher level of organization—an endeavor that lies beyond the possibilities of traditional modeling paradigms.

The lack of knowledge of quantitative biochemical details and the mathematical difficulties of accurately describing such complex systems of networked interactions necessitates the use of very simple models, in which certain abstractions have to be made, as well as aiming, in the first place, at learning about general principles. The hope behind this strategy is that some fundamental principles of regulation are typical properties of the systems under study and are robust to theoretical idealization, such that they are preserved in the simplifying model. In fact, it turned out that the idealization and simplifications are not only necessary but also sufficient for studying the generic, qualitative dynamics of a cellular regulatory system as a whole (rather than to predict specific quantitative responses).

A powerful model in this sense is the Boolean genetic networks in which time and values of activity levels of the genes are made discrete. Boolean networks were first proposed by Kauffman as a tool to address the fundamental problem of spontaneous generation of order in complex systems, in particular in metabolic and genetic networks (Kauffman, 1969, 1993). Here we will use the Boolean genetic network model as a simple modeling language to illustrate how the global dynamics of cell fate regulation can be understood as resulting from the interaction of signaling molecules organized as a network.

Sigmoidal Kinetics and Discretization of Molecular Interactions

In a Boolean network an element of the network represents a gene or a protein and can take only two values, ON (=1) and OFF (=0). The idealization of a gene or protein activity as an ON-OFF switch ("discretization") is based on the use of a step function to approximate a steep sigmoidal function (ultrasensitivity) for the function that maps the regulatory signal (input; e.g., level of a stimulating protein) to the response (output; e.g., the enzyme activity of the target) (Thomas et al., 1995).

However, the observed switchlike behavior can be more than just an approximation based on the limit of a steep sigmoidal input-output relation. From the analysis of nonlinear dynamic systems we know that sigmoidal functions can give rise to *bistability* in simple control circuits, for example, mutual inhibition between two proteins, A and B, given appropriate parameters. In such a feedback circuit only two states of the system will be stable, one in which protein A is low and B is high, and one in which A is high and B is low. Which state is taken by the system depends on the initial protein activities relative to threshold values (Glass and Kauffman, 1973).

The occurrence of sigmoidal input-output functions justifies the discretization. But why can we assume the presence of sigmoidal, ultrasensitivity functions for molecular interactions at all? Classical Michaelis-Menten enzyme kinetics gives rise to hyperbolic functions that exhibit saturation (flattening of the curve for high input signals) but not the sigmoid behavior at the low level of input signals necessary for bi- or multistability. The generation of sigmoidal threshold kinetics depends on how the underlying kinetic mechanism is capable of sensitivity amplification, and thus of generating ultrasensitivity (Koshland et al., 1982). The best-known situation that generates a sigmoidal behavior is cooperativity, as classically studied for multimeric proteins like hemoglobin. High cooperativity leads to ultrasensitivity. A cascade in a signaling pathway can also produce an all-or-nothing response, as has been shown for the MAP kinase signaling pathway (Ferrell and Machleder, 1998).

However, even without these rather complicated control structures it has been proposed that alone the particular nature of intracellular biochemical reactions, which involves reaction on surfaces rather than in free three-dimensional space (dimension restriction), and the relative stability of "transition complexes" (e.g., protein-protein complexes) contribute to a deviation from hyperbolic Michaelis-Menten kinetics toward sigmoidal kinetics (Savageau, 1995). Moreover, it has been suggested that stochastic fluctuations due to the low number of copies of reactants in the cell will promote sensitivity amplification (i.e., cause a gradual response to sharpen and behave like a threshold mechanism), even in the absence of more complicated control circuits (Paulsson et al., 2000).

The discretization now allows the use of logical functions as a simple way to directly capture the qualitative relationship between the inputs inter se and their joint effect on the target, for which information is readily available from common cell biology experiments. Molecular information processing that displays the basic features of logical gates in fact takes place at promoter elements or in multiprotein complexes (Yuh et al., 1997; Bray, 1995). Figure 8.3 shows two examples of the encoding of molecular interactions into Boolean functions.

A common criticism of this digitalization is that the assumption of ON/OFF states disregards the continuous nature of gene or protein levels in the cell. It should therefore be emphasized that the ON/OFF idealization is not an arbitrary coarsening, achieved by simply subdividing a gradual response into two qualities, nor does it represent an elementary molecular ON/OFF switch; rather, it reflects an ensemble behavior of molecules in biochemical reactions that gives rise to bistability. Therefore, an intermediate activity value of the enzymes around the threshold will be unstable, and the activity will move toward one of the two stable extremes, either maximal activity (corresponding to the ON state) or minimal activity (OFF state). The essence of the approximation thus lies in the equating of potentially differing threshold values across the network. Time discretization, on the other hand, reflects the two situations before and after a threshold has been crossed. Here the departure from reality lies in the synchrony of updating between all network elements and the simultaneous passage of multiple proteins through their threshold values within the same time interval.

Besides the theoretical justification, the discretization is also supported by the frequent experience of cell biologists that in *dose-response* experiments monitored at the *single-cell level*, the cellular response typically exhibits an all-or-none behavior (Ferrell and Machleder, 1998). Gradual responses, on the other hand, are mostly seen in "bulk" biochemical studies where the response is averaged over an ensemble of cells in which the respective threshold has a statistical distribution.

Basic Structure of Boolean Networks

We briefly review here the basic concepts of Boolean networks before discussing how they provide new insights into cell fate regulation. (For an introduction see also Somogyi and Sniegoski, 1996; Kaplan and Glass, 1997; Huang, 1999.) For the sake of simplicity and concreteness,

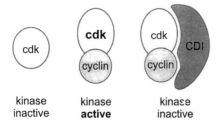

INPUTS		OUTPUT
Cyclin	CDI	cdk
0	0	0
0	1	0
1	0	1
1	1	0

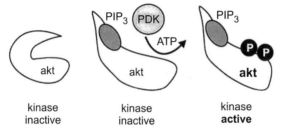

INPUTS		OUTPUT
PDK	PIP$_3$	akt
0	0	0
0	1	0
1	0	0
1	1	1

Figure 8.3 Encoding regulatory multiprotein complex protein interactions into simple two-input Boolean functions. *Upper panel*: cdk, the cyclin-dependent kinase (e.g., cdk4) is the target protein that receives, in this simplification, just two inputs from its upstream regulators, a cyclin (e.g., CD1) and a cdk-inhibitor, CDI (e.g., p27 or p21). The cdk requires association with a cyclin to be activated. The CDI binds to the cyclin-cdk complex (Guan et al., 1996) and inhibits cdk activity. This regulation can be encoded with a NOT IF Boolean function shown on the far right: The cdk is ON (output = 1) (active kinase) only *IF* CDI is *NOT* present (= 0) *AND* the cyclin is present (1).

Lower panel: The protein akt (= PKB) undergoes conformation change after binding of the phospholipid PIP$_3$, allowing the phosphorylation at two sites by the kinase PDK (PDK1 and PDK2) (Downward, 1998). The phosphorylated akt protein is now an active kinase (see also figure 8.1). This regulation represents a simple AND Boolean function: akt is active (output = 1) only if both PDK *AND* PIP$_3$ are present (= 1).

Note that in theses examples, CDI cannot bind cdk in the absence of cyclin, and PDK cannot act on akt before PIP$_3$ has bound. The Boolean function correctly encodes the net outcome of the interaction, but it does not capture these temporal subtleties, although the order of the inputs in the case of akt is essential for its activation by PDK.

let us assume a regulatory network that consists of protein-protein interactions (e.g., phosphorylation or proteolysis), as is the case for most interactions involved in controlling cell fate. As discussed above, the system can be abstracted as a network of N interconnected binary elements, $g_1, g_2, \ldots g_N$, representing proteins that at a given time t can take the values $ON: g_i(t) = 1$ or $OFF: g_i(t) = 0$. The ON value represents, for example, a kinase that is expressed and activated in a sufficient amount such that the total activity of all molecules is above the relevant threshold, whereas the OFF value represents the kinase whose activity is below the threshold or that is not expressed. Each network element g_i receives a number of k_i inputs, and is assigned a Boolean function that defines how the set of inputs affects the activity state of that element, which represents the output. Figure 8.4 shows an example of a network with $N = 4$ elements. For simplicity we restrict ourselves in this discussion to a subset of networks of low connectivity ($k \ll N$) and constant $k_i = k$ for all elements g_i.

A Boolean function represents the multiprotein complex in which specific protein-protein interactions determine the activity of the target protein (figure 8.3). The connections of the network are defined by the specificity of the interactions encoded in the molecular structure of the proteins (key-lock principle), and collectively specify a precise connection diagram of the network. The *network architecture* thus consists of the *topology* of the interactions (the wiring) and the *logical function* assigned to each network element (for instance, *"or," "and," "not if"* in figure 8.4). Since, in a first approximation, protein structure and its enzymatic function are determined by the gene sequence, the architecture of the network is "hardwired" in the genome and does not change unless affected by mutations (discussed later). What changes with time is the activity status of all network elements $g_i(t)$ at a given time, which collectively form a pattern that represents the *network state S* at that time point:

$$S(t) = [g_1(t), g_2(t) \ldots g_N(t)].$$

The Dynamic Behavior of the Network: Constraints Imposed by the Interactions

We are interested in the global dynamics of the regulatory network (i.e., in how the pattern of protein activity, or the network state $S(t)$, changes over time and governs cell behavior). Since the pattern of activity of all the proteins across the genome determines the cellular phenotype, every cell state corresponds to a distinct pattern of protein activity and, therefore, to a network state in the model. Thus, the dynamics of the network state determines the dynamics of the cell state, and hence of cell behavior and fate.

Network wiring diagram

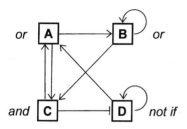

Protein activity state space

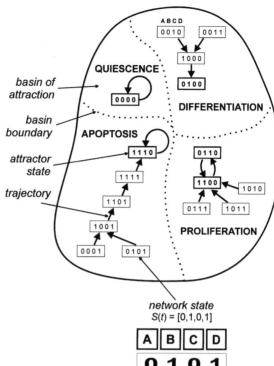

**Boolean functions
for A, B, C and D:**

A: " or "

INPUTS		OUTPUT
C	D	A
0	0	0
0	1	1
1	0	1
1	1	0

B: " or "

INPUTS		OUTPUT
A	B	B
0	0	0
0	1	1
1	0	1
1	1	0

C: " and "

INPUTS		OUTPUT
A	B	C
0	0	0
0	1	0
1	0	0
1	1	1

D: " not if "

INPUTS		OUTOUT
C	D	D
0	0	0
0	1	1
1	0	0
1	1	0

Figure 8.4 Basic principle of the Boolean network model and cellular states. An ($N = 4$ element) network is used as an example to illustrate how attractors arise from the interactions between the network elements. *Left:* network architecture of the four elements, A, B, C, and D, proteins of a regulatory network. In this idealized case, all elements have two inputs that can include input from the element itself, representing a direct feedback loop. The thin arrows indicate specific regulatory interactions. Each element is assigned a Boolean function, as shown in the corresponding boxes below.

Right: The state space and attractors, represented as proposed by Wuensche (1998). Each element can be ON (= 1) or OFF (= 0). The set of all the ON/OFF statuses for all elements defines a *network state* $S(t)$ for a given time point, depicted as a box with four binary digits. All possible network states together, in this case $2^4 = 16$, form the *state space*. Following the rules defined by the wiring diagram and the Boolean functions on the left, the individual network states transition into each other as indicated by the thick arrows, which represent the *trajectories*. Attractor states are in boldface. Trajectories can converge but not diverge, thereby contributing to the basins' draining into the attractors. The entire state space is divided into *basins of attraction* whose boundaries are shown with dotted lines. Each attractor and its basin can be equated to a cell fate: differentiation, quiescence, apoptosis, and proliferation. The proliferation attractor is a limit cycle (in this case containing two states) corresponding to the oscillatory gene expression patterns during repeated cell cycles.

Now, because of the interaction between the proteins, the network state can change only according to the wiring diagram and the logical rules for the interactions. For instance, if the wiring and rules are such that protein A is an input on protein B, and unconditionally inhibits protein B, then a network state with protein A being stably ON can only transition into the subset of network states where B is OFF. All network states in which both A and B are simultaneously ON would be logically forbidden (i.e., would correspond to unstable activity patterns that would transition into a stable one). Thus the dynamics of the network state is severely restricted. Therefore, cell behavior is channeled along the possibilities dictated by the wiring diagram. The constraints of global cell behavior reflect the constraints of the dynamics of the network state imposed by the network architecture. It is precisely in the analysis of these constraints that our chance lies to understand how cellular biochemistry, represented as a schematic diagram of signaling pathways, translates into a global, "observable" behavior of the cell.

State Space and Trajectories, Attractors and Basins of Attraction

An important concept for the description of network dynamics is the state space (figure 8.4). Briefly, in our binary approximation each network state $S(t)$ can be written as a string of length N of 1s and 0s, such as [111000111....]. A network state represents a point in the *state space*, the N-dimensional space that contains all possible network states. Figure 8.4 (right panel) shows a graphical representation of a state space of 16 states as a projection into two dimensions. Similar states that differ in the activity status of just a few proteins would be located close together in the state space (e.g., 111000 and 111001). The number of all possible states is immense because of combinatorial explosion: for a network of 10,000 distinct proteins (as estimated for a given mammalian cell), there would be a theoretical number of $2^{10000} \approx 10^{3000}$ different states. However, because of the restrictions discussed above, not all states are equally stable: a network state that is logically forbidden, and hence unstable, will transition to neighboring states when the Boolean function is executed.

Remember that the cellular regulatory network is an open system: under physiological conditions, the cell is submerged in a bath of free energy that is continuously replenished (e.g., available in the form of the molecule ATP, which is used for protein phosphorylation that drives the regulatory reactions represented by the Boolean functions). Thus, in a first approximation we can neglect energy considerations, and assume that execution of the Boolean functions for all the proteins of a network, also referred to as *updating* of the network, is therefore a spontaneous (exergonic) process under physiologic conditions, such that the network continuously updates itself. Consecutive updating of the network state causes the cellular regulatory network to travel along a chain of unstable states in the state space, thereby defining *trajectories* (figure 8.4).

Because of the logical constraints and the ensuing dynamics of the network state, the state space is not homogeneous: instead, unstable states are forced to migrate along trajectories into stable states as the Boolean rules are executed. Since the protein-protein interactions in the multiprotein complexes have unambiguous biochemical effects (which has allowed the representation as deterministic Boolean functions), a network state can have only one successor state while it can have multiple predecessor states. Thus, trajectories in the protein activity state space can converge but not diverge, and a network state does not have a memory of where it came from. This property is fundamental to cellular regulatory networks and gives rise to directionality and robustness of cellular behavior. In the absence of any regulatory interactions, all network states would be equally likely, and there would be no trajectories and driving force.

Since the state space is finite, trajectories will eventually hit upon states the network has already visited before, and because of the deterministic nature of updating, it will reenter the same succession of state transitions, thus forming loops in the state space. Of interest here are the cases where the loop of succeeding states consists of just a few states (a tiny fraction of the state space). In the extreme case, trajectories end at network states that, when updated, yield the same state (i.e., $S(t+1) = S(t)$). These *asymptomatic states*, which occupy either a small number of looping states in the state space or just one network state, are called the *attractor* states of the state space, since they attract the trajectories.[1] Thus, attractors can either be oscillatory (*limit-cycle attractors*) or stationary (*fixed-point attractor*). The small example network in figure 8.4 has one two-state limit cycle. The set of network states in the state space that will end up in the same attractor forms the *basin of attraction* of the respective attractor. *Basin boundaries* divide the state space into the basins of attraction. The state space thus has a structure: it is compartmentalized, and trajectories do not cross the basin boundaries. This imposes a restriction on the dynamics of the global protein activity patterns that is reflected in the rulelike behavior of the cell.

The restricted, attractor-driven dynamics of the network can be pictured as a marble on a landscape with valleys and pits that force the marble to follow a certain path that ends in one of the pits—these local minima would correspond to attractor states. The position of the marble at a given time represents a transient protein activity pattern, or network state, that displays an urge to move along a given trajectory. The choice of the destination attractor state depends on the marble's position (initial state), the "shape of the attractor landscape," and external perturbations. We now move on to confront the model's dynamics with experimental observations in real cells.

CELL STATES IN REAL CELLS AS ATTRACTORS

Since every cellular state corresponds to a distinct protein activity pattern, and the latter maps into a network state in the model, we have now placed cell fate dynamics into the conceptual framework of regulatory networks in which we have driving forces, directionality of processes, and stable, qualitatively distinct attractor states.

It has long been proposed that metabolic states or differentiated cell types represent stable, stationary states in genetic circuits that exhibit multistability or attractor states in genetic networks (Delbrück, 1949; Monod and Jakob, 1961; Kauffman, 1973; Thomas, 1998). In view of the enormous progress in our understanding of details of cell regulation in the past few years, we present here a refined picture of the meaning of attractor states as the most elementary emergent feature generated by the interaction of network elements. Based on the distinct dynamic features of cell fate regulation, we propose that *phenotypic cell states* (i.e., the various cell fates) are *attractor states of the underlying regulatory network* (Huang, 1999). In this model, the proliferating state, with its recurring succession of distinct protein activation patterns during repeated cell division cycles, would correspond precisely to a limit cycle attractor, while other cell states, such as differentiation, quiescence, or programmed cell death, would correspond to fixed-point attractors (figure 8.4, right side).

Equating real cellular states to attractor states of the network model allows us to address some aspects of real cell behavior from an integrative viewpoint, and to describe the relation between cell fate regulation and molecular pathways with a formal language instead of resorting to intuitive, ad hoc explanations based on arrow-arrow schemes (Figure 8.1).

We discuss in the following section some basic properties of real cell states in the light of the network model, and show how some phenomena that currently are described with an anthropomorphic verbiage, including terms like "decision," "switch of cell fates," "balance of signals," and the idea of "conflicting signals" (Raff, 1992; Evan and Littlewood, 1998), can be put on a formal foundation.

Discreteness and Mutual Exclusiveness of Cell Fate

In response to a set of heterogeneous and often opposing stimuli, cells have to make an all-or-nothing decision as to which fate to take: to divide, to differentiate, or to die. This has raised the question of how the cell shuts down, say, the multitude of biochemical pathways known to promote apoptosis when it "decides" to enter the growth state. The combination of cell states being attractors and the concept of attractor basins now predicts that cellular states are a priori mutually exclusive, since the network state can be in only one attractor at a time. This would also agree with the all-or-none nature of transitions between cell states and justify the use of the term "switch" to indicate transition between states. It has long been known that proliferation and differentiation are reciprocal (Goss, 1967). In fact, removing cells from the proliferating state by withdrawing the mitogenic factor from the medium not only arrests the cell cycle in pluripotent precursor cells but also very often triggers differentiation into the corresponding postmitotic mature cell, such as myocyte, adipocyte, or neuron (Olson, 1992; Harrison et al., 1985; Brüstle et al., 1999).

Even arresting the cell cycle by molecular manipulation of the regulatory network—for example, by overexpressing the cyclin-dependent kinase inhibitor p21 (a cell-cycle inhibitor)—induces differentiation in many cellular systems (Steinmann, 1994; Parker et al., 1995; Liu et al.,

1996; Erhardt and Pittman, 1998). In all these cases no differentiative instruction is given to the cell, just a signal that destabilizes the proliferative state, following which the network falls into the attractor of the differentiation state. In some cases, and depending on other factors (growth condition, cell type), a growth arrest can also result in activation of cell death.

Of logical necessity the regulatory network's wiring diagram not only establishes the attractor states but also determines the changes of the patterns of protein activity that occur in the cell as it travels along a trajectory into a given attractor state (figure 8.4). Such changes of the protein activity pattern then appear a posteriori in the traditional pathway-centered view as an amazingly well-orchestrated, target-oriented, and unlikely complicated process, when in fact these temporal changes of protein activity pattern are "hardwired" into the regulatory network that defines the basin of attraction and reflects the self-driven execution of the Boolean rules.

Robustness of Cell States

Equating cell states with attractor states implies that the trajectories of such states or the states themselves are intrinsically robust. Robustness is stability against a wide variety of perturbations. In simulation of networks, small perturbations consist of a "bit-flip" (changing the ON/OFF status of just a few proteins), and will kick the network into a nearby "forbidden" state. For minimal perturbations that hit only one network element at a time, the system will flow back along trajectories to the original attractor 80–90% of the time, as shown in computer simulations (Kauffman, 1993). This robustness represents an important buffer against biological noise, which is unavoidable, given the high local fluctuations of protein activity levels due to the spatially heterogeneous intracellular milieu with just a few thousands of copies of each distinct regulatory protein.

The observation that the same cell fate can be induced by treatment with different, unrelated chemicals—as impressively demonstrated by the ability of a variety of agents, including dimethylsulfoxide, retinoic acid, flavones, and benzene, among others, to induce the process of myeloid differentiation in promyelocytes (Birnie, 1988; Kalf and O'Connor, 1993)—supports the picture of robust attractors with converging trajectories.

Compartmentalization of the State Space: Classification of Environmental Inputs

Every major environmental input that perturbs the activity of one or more proteins will displace the network state from the current attractor state and place it at another spot somewhere in the state space, from which it will relax into the respective attractor. The state space is compartmentalized by the boundaries of basins of attraction. Thus, the cell classifies its environment into a few categories that correspond to the attractors, and maps every environmental perturbation of the network state into a distinct cell fate. This would explain why agents that do not bind to a specific receptor but instead directly affect the activity of many proteins, or why even physical perturbations can trigger a distinct cellular response as complex as differentiation, which is normally thought to be induced by a specific physiological signal and requires a concerted change of the activity of a large set of regulatory proteins.

The compartmentalization of the state space implies that the "decision" to take a cell fate is an inevitable process, and that the cell in a given attractor state has only a few possible choices. The state of undecidedness is dynamically unstable because a network state at the boundary between two attractor basins itself is unstable and has to move toward one of the two attractors. It is the protein activity pattern after the perturbation as such—no matter whether it was caused by one specific signal or multiple, disparate external signals—that will determine in which particular basin the cell will fall, since the activity profile as a whole determines the position of a network state relative to the basin boundaries. This explains why nominally opposing signals, such as the simultaneous stimulation to proliferate and to differentiate, still result in one distinct cell fate, such as apoptosis. The widely used pictures of "balance of signals" and "conflicting signals" inappropriately compress the multidimensionality of the decision-making process at the basin boundaries of the N-dimensional state space into a one-dimensional weighing event.

Physical Inputs and Attractor States

The self-organized nature of distinct cell states not only allows for nonbiological, "nonspecific" chemicals to elicit biologically coherent responses, but also brings cell fate regulation into the realm of physical influences present in the cell's environment. Experiments using micro-patterning technology to manipulate cell shape have shown that distinct cell fates can be induced by purely physical perturbations, such as change of cell shape and mechanical distortion of the cell (Chen et al., 1997; Huang and Ingber, 1999). These physical signals lack the ability to directly and specifically interact with molecular components of the regulatory machinery, although at some point they have to be transduced into a coherent protein activity pattern. The experimental finding that change of cell shape induces the same alterations of protein activity patterns as do specific growth factors provides experimental support for the idea that cell fates can be represented by attractor states of protein activity patterns (Huang et al., 1998; Huang and Ingber, 2000b).

The existence of attractors and their basins of attraction representing cell fates enables physical factors, such as cell geometry and distortion, to harness the molecular machinery of cell fate regulation. This will massively increase the odds for evolutionary linking of the intricate biochemistry of cell fate regulation with the physical world that lacks specific, molecularly encoded information, which may thereby have facilitated the evolution of larger organisms whose development and function are subject to the laws of macroscopic physics, yet are governed by genes and proteins (Huang and Ingber, 2000b).

CELL FATE REGULATION AS TRANSITIONS BETWEEN ATTRACTORS

How does the traditional description of cell fate regulation by serial molecular interactions, organized in signal transduction pathways, fit into the concept of network attractors? Cellular signaling triggered by the treatment with growth factors promotes the switching of the cell into another cell state, and hence must correspond to the switching of attractor states of the network. Therefore, a signaling cascade represents a perturbation that results in the transition between attractor states.

Signaling Events as Perturbation of Large Sets of Cellular Proteins

A transition between cell states requires a jump of the network state from one attractor state into the basin of another attractor state. Since attractor states are intrinsically robust, most such transitions will require the perturbation (change of activity status) of a large number of proteins. The network architecture determines size and composition of the set of network elements whose joint perturbation can cause a transition between attractor states, and which therefore form a *transition-causing set*. The optimal size of such transition-causing sets of perturbation targets can vary considerably.

This is illustrated in a network simulation, shown in figure 8.5, for a small regulatory network ($N = 12, k = 2$), in which perturbations of sets of m randomly chosen proteins were tested for their ability to cause the network to switch the attractor (measured as the number of occurrences of a specified transition). m is the size of the perturbation set, or the number of randomly chosen elements of the network that receive a bit-flip perturbation; thus, m can be thought of as *perturbation strength*.

The results show a rich, nonmonotonic behavior of transition frequencies as a function of the perturbation strength m. A high transition frequency for a given m indicates that many perturbation sets of size m exist that can cause that specified transition, and is a measure of "ease" and robustness of that transition. Thus, much as with attractor states, the transition-causing sets themselves can display a characteristic robustness. Each distinct transition (defined by origin and target attractor) exhibits a different pattern of robustness. For instance, some transitions occur rarely and only at perturbation strengths above a minimal value, while for other transitions there is an optimal perturbation strength.

This brings us back to the question of biochemical pleiotropy in signal transduction. The large number of proteins and genes whose activity is affected in the signal transduction cascades triggered by the activation of a cell surface receptor may precisely reflect the need for a perturbation set of a certain size in order to push the cell into another basin of attraction. Thus, during evolution of the regulatory network architecture, the "design" of signal transduction cascades and attractor landscape must be tightly coupled to produce networks in which a defined set of proteins needs to be turned ON/OFF to cause a certain transition between attractors. Such characteristic sets of perturbation targets can be likened to a code or password that provides access to a specific attractor from another specific attractor. The extent of biochemical pleiotropy in signal transduction thus might reflect evolution's fine-tuning of the m value, which in turn determines the

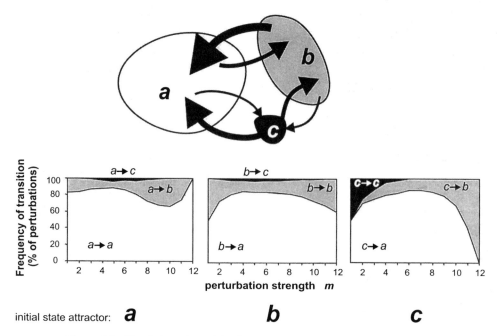

Figure 8.5 Simulation of transition between attractor states upon random perturbation. Results are for an example network of $N = 12$, $k = 2$ with a randomly determined architecture, leading to three attractor states, a, b, and c. The top panel provides a graphical representation of the three basins of attraction of different size (not drawn to scale). The arrows denote transitions, with thicker arrows indicating increased transition frequency (again not drawn to scale). The basins of attraction occupy varying fractions of the state space (of total $2^{12} = 4{,}096$ states): a, 83 percent; b, 15 percent; c, 2 percent. b is a limit cycle; a and c are fixed-point attractors. All three attractor states were used as initial states to be perturbed by randomly choosing a subset of m ($m = 1, 2, \cdots, N$) elements that will undergo a bit-flip. Per m value, 10,000 iterations of random selection and perturbation were performed, and the occurrence of the three possible transitions (including the ones to the same attractor) was measured. The lower panel shows, for all nine possible transitions from the three initial attractor states, the relative frequency of the three possible transitions as a function of the perturbation strength m. Note the nonmonotonic behavior: for instance, transition from a to b is most likely with perturbation sets of size $m = 10$, and least likely with $m = 5$.

constraint or ease for evolving physiological fate switching signals that will "hook up" the network to extracellular cues. The information-coding capacity of transition-causing sets allows the genome to encode under which specific circumstances a cell can switch between which cell states.

The restriction imposed on transition events, as manifested in the requirement for a pleiotropic response to signals, solves a fundamental dilemma of systems with multiple attractor states that are all intrinsically stable. Without barriers to transitions, in the noisy cellular microenvironment the cell would randomly bounce between the attracting states. Too robust attractors would prevent any change of cell state. The "code-dependent" restriction of transitions allows the regulatory network to maintain cell fates as robust, self-stabilizing entities in the form of attractors, and yet to allow (tightly controlled) switches between them.

Thus, the architecture of the regulatory network is poised to unite local stability and global flexibility. Specifically, the transitions from the differentiation or quiescence attractor into the growth or the apoptosis attractor can so be restricted to a narrow range of specific external conditions. In this sense, the biochemical pleiotropy acts as an epigenetic safeguard against uncontrolled growth or death. Pleiotropy is a blessing for tightness of regulation, not a harm to signal specificity.

Analysis of gene expression with microarrays reveals groups of "coregulated" genes, which form regulons. As discussed in chapter 6, such sets of genes have common upstream regulatory motifs. It is often simply assumed that such genes "share the same or a similar function." One reason why a single "function" is controlled by a whole group of genes that are hardwired by the genomic sequence to be coregulated might be precisely the necessity for having jointly acting genes (transition-causing sets) to trigger phenotypic changes of the cell. For instance, stimulating fibroblasts with growth factors to undergo the transition from the quiescence state to the growth state activates a set of over 60 genes (Fambrough et al., 1999). This set of "immediate early response genes" might represent the specific transition-causing perturbation set that encodes the key to the growth state. In fact, activation of growth via the bFGF receptor or the PDGF receptor elicits almost an identical set of genes, although the immediate postreceptor biochemical events for these two growth factors are different.

Overlapping Perturbation Sets, Multiple-function Proteins, and Master Proteins

Since a given transition requires the activity change of a large, defined set of proteins, control of cell fate is distributed among a large number of proteins. The profile of protein presence and activity of the cell determines whether a given transition-causing set can be recruited. Thus, the effect of a molecular signal is a function of the "cellular context" defined by culture conditions, cell type, treatments, and so on. In network simulations, signaling proteins that are part of more than one (different) transition-causing set are readily found, indicating that transition-causing sets often overlap. This translates directly into the notion that regulatory proteins, such as Ras and Myc, can be part of different perturbation sets that cause different transitions. This in turn would explain the frequent but counterintuitive observation that signal transduction proteins can have opposing effects on cell fate—depending on the cellular context (figure 8.2).

The proteins that are part of a transition-causing set can be viewed as forming a rather loose functional module. The concerted activation of such a set would be achieved most efficiently by a wiring diagram in which a "master protein" would have as its downstream targets precisely such a set of proteins (or a portion of it). Cell surface receptors obviously are such master switches. Inside the cell the Ras protein, the ERK kinases, PI3K, or the Myc transcription factor might represent such master proteins that have downstream effects that ramify broadly, yet serve the same "function," as represented in the concept of regulons (Tavazoie et al., 1999). This gives rise to the appearance of functional modules that appear to be "prewired" for switching cell states and would explain the evolutionary conservation of entire pathways from yeast to mammals, such as the MAPK- cascades, although they serve different "functions" in these species (Waskiewicz and Cooper, 1995).

Environmental Cues: Epigenetic Modification of Attractor Landscape and Pseudo Attractors

Imagine now that in the model network, one would keep some network elements chronically in a given activity status, say ON. Although this can be seen as a change of the cellular context discussed above, we can express such changes in a more formal way. A persistent fixation of the value of some network elements will obviously alter the attractor landscape to some extent. Some attractor basins could shrink or even disappear, and others could enlarge. The requirement for a transition between attractors could be affected (i.e., the accessibility to certain attractors might increase or decrease, or the "access code" to a specific attractor could become "invalid"). New attractors could be created if the chronic perturbations affect certain network elements. Thus, the network architecture, inscribed in the genomic sequence, is not the sole determinant of the "effective" structure of the attractor landscape; (reversible) environmental factors help in the fine sculpting of its final shape.

The attractor landscape undergoes *epigenetic modification*. To distinguish the *effective* attractors that depend on continuous, external influence from the "true," *nominal* attractors determined by the regulatory network's hardwired architecture, we use the term *pseudo attractor*. This term should imply that these attractors exist only as long as the external cues persist, and are not an intrinsic property of the network. On the other hand, the genome can express the hardwired, nominal attractors only if all network elements are free to flip between ON and OFF.

The concept of pseudo attractors in regulatory networks is important because real cells are also exposed to continuous signals from their tissue environment, such as those originating from paracrine and juxtacrine (cell-cell) interactions or from interaction with the extracellular matrix. These contextual influences are thought to account for many of the characteristic features of the differentiated phenotype that are often lost when cells are removed from their natural tissue microenvironment and grown in culture.

The picture of a (pseudo) attractor landscape allows formulation of the biochemical effects of a regulatory input in a more general way. While we have discussed the capability of signaling molecules to switch cell fate, the epigenetic modification of the attractor landscape as it is defined in the network architecture stored in the genome will also influence the capacity of the external signals to switch cell state. This brings together genetic determination and environmental modulation. In a more encompassing category of thinking, one could envisage using the idea of pseudoattractor as a conceptual framework for describing the riddle of the relationship among genes, environment, and noise in determining the phenotype of an organism.

EXTENSION OF THE MODEL: CELL TYPES AND CANCER

Since we are focusing on cell fate regulation as the basis for tissue development and homeostasis, we have so far discussed the regulation within one tissue (i.e., within just one cell type) by equating attractor states with the various states that a cell can take. When the whole organism is considered, we have to deal with different cell types and the development of pluripotent cells into variously differentiated cells. As mentioned earlier, it has long been suggested that differentiated cell types correspond to attractor states. Kauffman refined this idea and placed it into the framework of genetic Boolean networks (Kauffman, 1973; 1993). He went further proposing that a cancer cell can be equated with a cell type: one that is abnormal and corresponds to an attractor that is normally not visited by the organism. We briefly discuss here how the model of cell fate regulation presented in this chapter can be extended to the whole-organism level in order to embrace the diversity of cell types and the origin of cancer, which, as alluded to in the introduction to this chapter, can be triggered by a disruption of cell fate regulation.

Cell Types as Sets of Attractors

Since every cell type has a distinct (although overlapping) protein activity pattern and can undergo the analogous cell fates, the extended model would be that each tissue occupies a region of the protein activity state space that consists of several attractors representing the various alternative cell states of that cell type. The growth attractor would correspond to the self-renewing stem cell of that tissue, which upon an appropriate stimulus can transition either into a differentiation attractor, representing a terminally differentiated cell, or into a quiescence attractor, representing the "dormant," pluripotent state characteristic of stem cells (Fuchs and Segre, 2000). An important aspect of this view is the varying accessibility to the cell state attractors in different tissues, which is the basis for the differences of tissue dynamics. For example, the terminally differentiated neural tissues have large differentiation attractors but normally almost inaccessible growth or apoptosis attractors, whereas in intestinal, blood, and skin cells, which represent high turnover, self-renewing tissues, the latter attractors are large and easily accessible.

Since many proteins are expressed only during embryonic development and organogenesis, one will have to assume that in the entire genomic protein activity state space, regions exist with attractors that correspond to embryonic cell types not present in the mature organism. These attractors are occupied only during development, and are no longer accessible due to a different set of tissue cues in the adult organism that, via epigenetic modification of the attractor landscape, establish barriers that prevent access to the embryonic attractors. These "archaic attractors," however, are still lurking somewhere in the state space and might play a role in the origin of tumors.

Cancer: Reactivation of Embryonic Attractors and Enlargement of the Basin of the Growth Attractor

In contrast to the epigenetic cues from the environment discussed so far, mutations cause permanent changes to the wiring diagram of the genome: they can affect the activity of a regulatory protein (permanently shutting it off or turning it on) or its interaction properties, (e.g., disrupt an interaction or change target specificity). This rewiring of the network can result in the permanent distortion of the attractor landscape such that novel attractors are created or existing ones that are inaccessible in a given adult tissue become accessible in response to an appropriate set of external stimuli. Based on this model, tumorigenic mutations would lead to an enlargement of the basin of the growth attractor in which the cell would be trapped. This increased robustness of the growth state would result in cell proliferation in a broader range of conditions (up to full independence of mitogenic signals), thus contributing to cellular fitness in the microevolutionary development of tumors. The most likely way for mutations to achieve an enlarged growth attractor would be to co-opt a nearby "archaic attractor" representing an embryonic version of the same cell type, by making it (or a distorted version of it) accessible from within an adult cell type.

Unlike cells in the tissue of mature organisms, in which the differentiation attractor is dominant such that proliferating cells are rare, in the embryo bulk cell proliferation is required for tissue expansion; thus this embryonic attractor would be expected to be a large and robust growth attractor. In fact, hallmarks of tumor cells are not only the increased stability of the proliferative state, as manifested in the reduced requirement for growth factors to proliferate, but also the features of immaturity in terms of differentiation characteristics. Tumor cells often express embryonic proteins, such as alpha-fetoprotein, placental alkaline phosphatase, or carcino-embryonic protein (Jacobs and Haskell, 1991). Moreover, phenotypes such as cell migration, tissue invasion, and stimulation of blood vessel growth, which contribute to the malignancy of tumor cells, are all normal features of embryonic cells. Numerous other generic properties of malignant cells—such as the initially increased tendency to apoptosis, the occurrence of trans-differentiation, and the existence of a few but distinct subtypes of a given cancer type—are consistent with the concept of attractors, but are beyond the scope of this chapter.

CONCLUSION AND OUTLOOK

In the wake of the unprecedented success of genomic technologies, scholars of biology have criticized the reductionist nature of "big biology" and its use of large-scale, high-throughput procedures to dissect and discover—rather than to understand. They have advocated an integrative view of living systems, in particular through the optics of the emerging science of complex systems (Strohman, 1997; Coffey, 1998; Rose, 1998; Lewontin, 2000; Huang, 2000).

As a concrete case of such an integrative approach, we have focused in this chapter on the problem of cell fate regulation in mammals as first level of integration of gene effects into a higher-level functionality. We have introduced the established ideas of Boolean networks as a generic model for signal transduction in cell fate regulation, and have demonstrated how the collective behavior of the members of an interaction network can give rise to a rich and complex, but ordered, behavior restricted by rules.

However, we need to stress here that the Boolean network model is no doubt an oversimplification, more of a symbolic modeling language that serves as a tool to study—with the necessary coarse graining and abstraction—the behavior of large systems of interacting molecules in an integrated manner. Modifications and refinements of the theoretical model system will be necessary to consider additional features of biochemical reality, such as the temporal succession of inputs (including differential affinity of interactions), asynchrony of network updating (including differential kinetics), and multiple thresholds for enzymatic activities.

We have also treated cells as homogeneous entities, though it is increasingly appreciated that signal transduction takes place on subcellular compartments that involve scaffold proteins (Tsunoda et al., 1998; Burack and Shaw, 2000). The spatial organization of the interacting partners could be encoded in more complex Boolean functions.

Moreover, to achieve a complete picture of regulatory networks, one will have to extend the model beyond the single layer of a protein interaction network and include the protein-gene network (which comprises the interaction between transcription factors and cis-regulatory elements) for which gene expression profiling is already providing experimental data (Tavazoie et al., 1999; Bucher, 1999).

The discrete network model should by no means obviate the need for a detailed quantitative analysis of molecular machineries of cell regulation. However, Boolean networks are at the moment one of the simplest formalisms that can capture a wide range of properties of cell regulation into a coherent, formal framework that does justice to our mostly qualitative knowledge of molecular interactions, and, as such, can reveal some basic principles for a truly integrative approach in postgenomic biology.

It is obvious that beyond the generic model with anonymous genes and proteins, we will have to incorporate the specifics of existing molecular networks into our model. The spate of gene-specific data triggered by recent advances in genomic technologies, notably DNA microarray-based gene expression profiling (Young, 2000; van Hal et al., 2000) and proteomics (Dutt and Lee, 2000), hold promise of enabling this. However, profiling data are currently analyzed at a descriptive level (e.g., by clustering of genes or conditions based on similarities of expression patterns).

A first step beyond the descriptive interpretation of such functional profile data, although still in the embryonic stage, is the efforts to "reverse engineer" the complete network architecture based on gene expression profiles. Algorithms have been proposed that—at least in simple, idealized theoretical network models—can infer from series of gene activation profiles on to the underlying network architecture (Liang et al., 1998; Akutsu et al., 2000). Perhaps more practicable are the recently initiated collaborative efforts to build signal transduction pathway databases based on published, experimentally confirmed molecular interactions of the regulatory network (e.g., Biopathways Consortium and Cell Signaling Networks Database). Such electronic databases will ultimately lead to a map of the complete wiring architecture of regulatory networks of mammalian cells, similar to the classical charts of metabolic networks. These regulatory maps should serve to organize and graphically represent the accumulated knowledge of pathways and their relationships, and can also teach valuable lessons about universal, large-scale design principles of biological networks (Jeong et al., 2000).

However, such knowledge databases of regulatory networks will be mere static graphs of pathways if one does not breathe life into these maps by *simulating* their dynamic behavior, as demonstrated in this chapter with theoretical, anonymous networks. Given the intrinsic robustness of biological networks (Alon et al., 1999; Salazar-Ciudad et al., 2000), it is likely that an exhaustive knowledge of all the quantitative parameters might not be necessary. Validation of the network structure will be possible by comparing the simulated dynamic behavior with real data obtained from specific perturbation experiments in which, using current and future technologies, gene expression profiles (transcriptome), and protein expression and activity profiles (proteome, phosphorylome, activome, etc.), are monitored (MacBeath and Schreiber, 2000).

This should launch the next big (postgenomic) endeavor of analyzing *the structure of the gene activity state space* of real cellular systems, mapping the real attractor landscape of the human genome, and studying its biological significance. The information on the dynamics might be more valuable than the static network architecture map, since it links directly to global cellular behavior. Knowledge of the state space structure could pave the way toward developing the predictive capabilities in cell fate regulation that one will need to understand regulative disorders, such as cancer, and to explore the differentiation potential (neighboring, accessible attractors) of somatic stem cells in order to design therapeutics for a wide variety of diseases.

NOTE

1. Having few attractors that are either fixed points or small cycles is a property that belongs to only a subset of possible network architectures, namely those that are sparsely connected ($k \ll N$), as is the case in real cellular regulatory networks, and exhibit a bias toward a subset of Boolean functions that is most likely to be implemented by molecular interactions (high internal homogeneity). (For details see Kauffman, 1993.) Hereafter we will deal with this subset of biologically reasonable network architectures.

REFERENCES

Akutsu, T., Miyano, S., and Kuhara, S. (2000). Algorithms for inferring qualitative models of biological networks. *Pac. Symp. Biocomput.* 293–304.

Alon, U., Surette, M. G., Barkai, N., and Leibler, S. (1999). Robustness in bacterial chemotaxis. *Nature* 397: 168–171.

Barbacid, M. (1987). Ras genes. *Annu. Rev. Biochem.* 56: 779–827.

Birnie, G. D. (1988). The HL60 cell line: A model system for studying human myeloid cell differentiation. *Br. J. Cancer* 9: 41–45.

Bray, D. (1995). Protein molecules as computational elements in living cells. *Nature* 376: 307–312.

Brunet, A., and Pouysségur, J. (1997). Mammalian MAP kinase modules: How to transduce specific signals. *Essays Biochem.* 32: 1–16.

Brustle, O., Jones, K. N., Learish, R. D., Karram, K., Choudhary, K., Wiestler, O. D., Duncan, I. D., and McKay, R. D. (1999). Embryonic stem cell-derived glial precursors: A source of myelinating transplants. *Science* 285: 754–756.

Bucher, P. (1999). Regulatory elements and expression profiles. *Curr. Opinion Struct. Biol.* 9: 400–407.

Burack, W. R., and Shaw, A. S. (2000). Signal transduction: Hanging on a scaffold. *Curr. Opinion Cell Biol.* 12: 211–216.

Chao, M. V. (1992). Growth factor signaling: Where is the specificity? *Cell* 68: 995–997.

Chen, C. S., Mrksich, M., Huang, S., Whitesides, G. M., and Ingber, D. E. (1997). Geometric control of cell life and death. *Science* 276: 1425–1428.

Chervitz, S. A., Aravind, L., Sherlock, G., Ball, C. A., Koonin, E. V., Dwight, S. S., Harris, M. A., Dolinski, K., Mohr, S., Smith, T., Weng, S., Cherry, J. M., and Botstein, D. (1998). Comparison of the complete protein sets of worm and yeast: Orthology and divergence. *Science* 282: 2022–2028.

Coffey, D. S. (1998). Self-organization, complexity and chaos: The new biology for medicine. *Nat. Med.* 4: 882–885.

Coller, H. A., Grandori, C., Tamayo, P., Colbert, T., Lander, E. S., Eisenman, R. N., and Golub, T. R. (2000). Expression analysis with oligonucleotide microarrays reveals that MYC regulates genes involved in growth, cell cycle, signaling, and adhesion. *Proc. Natl. Acad. Sci. USA* 97: 3260–3265.

Constantinou, A., and Huberman, E. (1995). Genistein as an inducer of tumor cell differentiation: Possible mechanisms of action. *Proc. Soc. Exp. Biol. Med.* 208: 109–115.

Delbrück, M. (1949). Discussion. In CNRS (ed.), *Colloques internationaux du Centre national de la recherche scientifique. Unités biologiques douées de continuité génétique.* Paris, CNRS.

Di Cunto, F., Topley, G., Calautti, E., Hsiao, J., Ong, L., Seth, P. K., and Dotto, G. P. (1998). Inhibitory function of p21Cip1/WAF1 in differentiation of primary mouse keratinocytes independent of cell cycle control. *Science* 280: 1069–1072.

Downward, J. (1998). Lipid-regulated kinases: Some common themes at last. *Science* 279: 673–674.

Dutt, M. J., and Lee, K. H. (2000). Proteomic analysis. *Curr. Opinion Biotechnol.* 11: 176–179.

Endy, D., and Brent, R. (2001). Modelling cellular behaviour. *Nature* 409 (suppl.): 391–395.

Erhardt, J. A., and Pittman, R. N. (1998). Ectopic p21(WAF1) expression induces differentiation-specific cell cycle changes in PC12 cells characteristic of nerve growth factor treatment. *J. Biol. Chem.* 273: 23517–23523.

Evan, G., and Littlewood, T. (1998). A matter of life and cell death. *Science* 281: 1317–1322.

Fambrough, D., McClure, K., Kazlauskas, A., and Lander, E. S. (1999). Diverse signaling pathways activated by growth factor receptors induce broadly overlapping, rather than independent, sets of genes. *Cell* 97: 727–741.

Ferrell, J. E., Jr., and Machleder, E. M. (1998). The biochemical basis of an all-or-none cell fate switch in *Xenopus* oocytes. *Science* 280: 895–898.

Franke, T., Kaplan, D. R., Cantley, L. C., and Toker, A. (1997). Direct regulation of the akt proto-oncogene product by phosphatidylinositol-3,4-bisphosphate. *Science* 275: 665–668.

Fuchs, E., and Segre, J. A. (2000). Stem cells: A new lease on life. *Cell* 100: 143–155.

Gire, V., Marshall, C., and Wynford-Thomas, D. (2000). PI-3-kinase is an essential anti-apoptotic effector in the proliferative response of primary human epithelial cells to mutant RAS. *Oncogene* 19: 2269–22676.

Glass, L., and Kauffman, S. A. (1973). The logical analysis of continuous, non-linear biochemical control networks. *J. Theor. Biol.* 39: 103–129.

Goss, R. J. (1967). The strategy of growth. In H. Teir and T. Rytömaa (eds.), *Control of Cellular Growth in the Adult Organism.* London: Academic Press.

Guan, K. L., Jenkins, C. W., Li, Y., O'Keefe, C. L., Noh, S., Wu, X., Zariwala, M., Matera, A. G., and Xiong, Y. (1996). Isolation and characterization of p19INK4d, a p16-related inhibitor specific to CDK6 and CDK4. *Mol. Biol. Cell.* 7: 57–70.

Harrison, J. J., Soudry, E., and Sager, R. (1985). Adipocyte conversion of CHEF cells in serum-free medium. *J. Cell. Biol.* 100: 429–434.

Hartwell, L. H., Hopfield, J. J., Leibler, S., and Murray, A. W. (1999). From molecular to modular cell biology. *Nature* 402(6761 suppl): C47–C52.

Huang, S. (1999). Gene expression profiling, genetic networks and cellular states: An integrating concept for tumorigenesis and drug discovery. *J. Mol. Med.* 77: 469–480.

Huang, S. (2000). The practical problems of post-genomic biology. *Nat. Biotechnol.* 18: 471–472.

Huang, S., Chen, S. C., Whitesides, G. M., and Ingber, D. E. (1998). Cell-shape-dependent control of p27Kip and cell cycle progression in human capillary endothelial cells. *Mol. Biol. Cell* 9: 3179–3193.

Huang, S., and Ingber, D. E. (1999). The structural and mechanical complexity of cell-growth control. *Nat. Cell. Biol.* 1: E131–E138.

Huang, S., and Ingber, D. E. (2000a). Shape-dependent control of cell growth, differentiation and apoptosis: Switching between attractors in cell regulatory networks. *Exp. Cell. Res.* 261: 91–103.

Huang, S., and Ingber, D. E. (2000b). Regulation of cell cycle and gene activity patterns by cell shape: Evidence for attractors in real regulatory networks and the selective mode of cellular control. *InterJ. Genet.* 238.

Jacobs, E. L., and Haskell, C. M. (1991). Clinical use of tumor markers in oncology. *Curr. Problems Cancer* 15: 299–360.

Jeong, H., Tombor, B., Albert, R., Oltvai, Z. N., and Barabasi, A. L. (2000). The large-scale organization of metabolic networks. *Nature* 407: 651–654.

Kalf, G. F., and O'Connor, A. (1993). The effects of benzene and hydroquinone on myeloid differentiation of HL-60 promyelocytic leukemia cells. *Leukemig Lymphoma* 11: 331–338.

Kaplan, D., and Glass, L. (1995). *Understanding Nonlinear Dynamics.* New York: Springer-Verlag.

Kauffman, S. A. (1969). Metabolic stability and epigenesis in randomly constructed genetic nets. *J. Theor. Biol.* 22: 437–467.

Kauffman, S. A. (1973). Control circuits for determination and transdetermination. *Science* 181: 310–318.

Kauffman, S. A. (1993). *The Origins of Order.* New York: Oxford University Press.

Kauffmann-Zeh, A., Rodriguez-Viciana, P., Ulrich, E., Gilbert, C., Coffer, P., Downward, J., and Evan, G. (1997). Suppression of c-Myc-induced apoptosis by Ras signalling through PI(3)K and PKB. *Nature* 385: 544–548.

Keverne, E. B. (1997). An evaluation of what the mouse knockout experiments are telling us about mammalian behaviour. *Bioessays* 19: 1091–1098.

Knudson, A. G. (1996). Hereditary cancer: Two hits revisited. *J. Cancer Res. Clin. Oncol.* 122: 135–140.

Koshland, D. E., Goldbeter, A., and Stock, J. B. (1982). Amplification and adaptation in regulatory and sensory systems. *Science* 217: 220–225.

Krasilnikov, M. A. (2000). Phosphatidylinositol-3 kinase dependent pathways: The role in control of cell growth, survival, and malignant transformation. *Biochemistry* (Moscow) 65: 59–67.

Kulyk, W. M., and Hoffman, L. M. (1996). Ethanol exposure stimulates cartilage differentiation by embryonic limb mesenchyme cells. *Exp. Cell. Res.* 223: 290–300.

Lavoie, J. N., L'Allemain, G., Brunet, A., Müller, R., and Pouysségur, J. (1996). Cyclin D1 expression is regulated positively by the p42/p44MAPK and negatively by the p38/HOGMAPK pathway. *J. Biol. Chem.* 271: 20608–20616.

Lewontin, R. C. (2000). *The Triple Helix: Gene, Organism, and Environment.* Cambridge, Mass.: Harvard University Press.

Liang, S., Fuhrmann, S., and Somogyi, R. (1998). REVEAL, a general reverse engineering algorithm for inference of genetic network architecture. *Pac. Symp. Biocomput.* 3: 18–29.

Liu, M., Lee, M. H., Cohen, M., Bommakanti, M., and Freedman, L. P. (1996). Transcriptional activation of the Cdk inhibitor p21 by vitamin D3 leads to the induced differentiation of the myelomonocytic cell line U937. *Genes Dev.* 10: 142–153.

MacBeath, G., and Schreiber, S. L. (2000). Printing proteins as microarrays for high-throughput function determination. *Science* 289: 1760–1763.

Mayo, M. W., Wang, C. Y., Cogswell, P. C., Rogers-Graham, K. S., Lowe, S. W., Der, C. J., and Baldwin, A. S. Jr. (1997). Requirement of NF-kappaB activation to suppress p53-independent apoptosis induced by oncogenic Ras. *Science* 278: 1812–1815.

McAdams, H. H., and Arkin, A. (1998). Simulation of prokaryotic genetic circuits. *Annu. Rev. Biopohys. Biomol. Struct.* 27: 199–224.

Melton, D. W. (1994). Gene targeting in the mouse. *Bioessays* 16: 633–638.

Monod, J., and Jakob, F. (1961). General conclusions: Teleonomic mechanisms in cellular metabolism, growth and differentiation. *Cold Spring Harbor Symp. Quant. Biol.* 26: 389–401.

Mooney, D. J., Hansen, L. K., Vacanti, J. P., Langer, R., Farmer, S. R., and Ingber, D. E. (1992). Switching from differentiation to growth in hepatocytes: Control by extracellular matrix. *J. Cell. Phys.* 151: 497–505.

Olson, E. N. (1992). Interplay between proliferation and differentiation within the myogenic lineage. *Dev. Biol.* 154: 261–272.

Pardee, A. B. (1989). G1 events and regulation of cell proliferation. *Science* 246: 603–608.

Parker, S. B., Eichele G., Zhang P., Rawls A., Sands A. T., Bradley A., Olson E. N., Harper J. W., and Elledge S. J. (1995). p53-independent expression of p21Cip1 in muscle and other terminally differentiating cells. *Science* 267: 1024–1027.

Paulsson, J., Berg, O. G., and Ehrenberg, M. (2000). Stochastic focusing: Fluctuation-enhanced sensitivity of intracellular regulation. *Proc. Natl. Acad. Sci. USA* 97: 7148–7153.

Pawson, T., and Scott, J. D. (1997). Signaling through scaffold, anchoring, and adaptor proteins. *Science* 278: 2075–2080.

Phillips, R. L., Ernst, R. E., Brunk, B., Ivanova, N., Mahan, M. A., Deanehan, J. K., Moore, K. A., Overton, G. C., and Lemischka, I. R. (2000). The genetic program of hematopoietic stem cells. *Science* 288: 1635–1640.

Raff, M. C. (1992). Social controls on cell survival and cell death. *Nature* 356: 397–399.

Reznikoff, W. S. (1992). The lactose operon-controlling elements: A complex paradigm. *Mol. Microbiol.* 6: 2419–2422.

Rose, S. (1998). *Lifelines: Biology Beyond Determinism.* Oxford: Oxford University Press.

Rovera, G., O'Brien, T. G., and Diamond, L. (1979). Induction of differentiation in human promyelocytic leukemia cells by tumor promoters. *Science* 204: 868–870.

Salazar-Ciudad, I., Garcia-Fernandez, J., and Sole, R. V. (2000). Gene networks capable of pattern formation: From induction to reaction-diffusion. *J. Theor. Biol.* 205: 587–603.

Savageau, M. A. (1995). Michaelis-Menten mechanism reconsidered: Implications of fractal kinetics. *J. Theor. Biol.* 176: 115–124.

Sawyers, C. L., Denny, C. T., and Witte, O. N. (1991). Leukemia and the disruption of normal hematopoiesis. *Cell* 64: 337–350.

Serrano, M., Lin, A. W., McCurrach, M. E., Beach, D., and Lowe, S. W. (1997). Oncogenic ras provokes premature cell senescence associated with accumulation of p53 and p16 INK4a. *Cell* 88: 593–602.

Shastry, B. S. (1994). More to learn from gene knockouts. *Mol. Cell Biochem.* 136: 171–182.

Somogyi, R., and Sniegoski, C. A. (1996). Modelling the complexity of genetic networks: Understanding multigenic and pleiotropic regulation. *Complexity* 1: 45–63.

Steinman, R. A., Hoffman, B., Iro, A., Guillouf, C., Liebermann, D. A., and el-Houseini, M. E. (1994). Induction of p21 (WAF-1/CIP1) during differentiation. *Oncogene* 9: 3389–3396.

Strohman, R. C. (1997). The coming Kuhnian revolution in biology. *Nat. Biotechnol.* 15: 194–200.

Tan, P. B., and Kim, S. K. (1999). Signaling specificity: The RTK/RAS/MAP kinase pathway in metazoans. *Trends Genet.* 15: 145–149.

Tavazoie, S., Hughes, J. D., Campbell, M. J., Cho, R. J., and Church, G. M. (1999). Systematic determination of genetic network architecture. *Nat. Genetics* 22: 281–285.

Taylor, S. J., and Shalloway, D. (1996). Cell cycle-dependent activation of Ras. *Curr. Biol.* 6: 1621–1627.

Thomas, R. (1998). Laws for the dynamics of regulatory networks. *Int. J. Dev. Biol.* 42 (3, spec no): 479–485.

Thomas, R., Thieffry, D., and Kaufman, M. (1995). Dynamical behaviour of biological regulatory networks. I. Biological role of feedback loops and practical use of the concept of loop-characteristic state. *Bull. Math. Biol.* 57: 247–276.

Tsunoda, S., Sierralta, J., and Zuker, C. S. (1998). Specificity in signaling pathways: Assembly into multimolecular signaling complexes. *Curr. Opinion Genet. Dev.* 8: 419–422.

Tyson, J. J. (1999). Models of cell cycle control in eukaryotes. *J. Biotechnol.* 71: 239–244.

van Hal, N. L., Vorst, O., van Houwelingen, A. M., Kok, E. J., Peijnenburg, A., Aharoni, A., van Tunen, A. J., and Keijer, J. (2000). The application of DNA microarrays in gene expression analysis. *J. Biotechnol.* 78: 271–280.

Venter, J. C., Adams, M. D., Myers, E. W., et al. (2001). The sequence of the human genome. *Science* 291: 1304–1351.

Waskiewicz, A. J., and Cooper, J. A. (1995). Mitogen and stress response pathways: MAP kinase cascades and phosphatase regulation in mammals and yeast. *Curr. Opinion Cell. Biol.* 7: 798–805.

Weinberg, R. A. (1989). Oncogenes, antioncogenes, and the molecular bases of multistep carcinogenesis. *Cancer Res.* 49: 3713–3721.

Weng, G., Bhalla, U. S., and Iyengar, R. (1999). Complexity in biological signaling systems. *Science* 284: 92–96.

Woodbury, D., Schwarz, E. J., Prockop, D. J., and Black, I. B. (2000). Adult rat and human bone marrow stromal cells differentiate into neurons. *J. Neurosci. Res.* 61: 364–370.

Wuensche, A. (1998). Genomic regulation modeled as a network with basins of attraction. *Pac. Symp. Biocomput.* 3: 89–102.

Young, R. A. (2000). Biomedical discovery with DNA arrays. *Cell* 102: 9–15.

Yu, Z. W., and Quinn, P. J. (1994). Dimethyl sulphoxide: A review of its applications in cell biology. *Biosci. Rep.* 14: 259–281.

Yuh, C.-H., Bolouri, H., and Davidson, E. H. (1997). Genomic cis-regulatory logic: Experimental and computational analysis of a sea urchin gene. *Science* 279: 1896–1902.

SUGGESTED READINGS

Huang, S. (1999). Gene expression profiling, genetic networks and cellular states: An integrating concept for tumorigenesis and drug discovery. *J. Mol. Med.* 77: 469–480.

Lewontin, R. C. (2000). *The Triple Helix: Gene, Organism, and Environment.* Cambridge, Mass.: Harvard University Press. An elegantly written, concise, and accessible text that addresses the limits of the gene-centered view of modern molecular biology, and confronts it with the questions concerning the relationship between genotype and phenotype, genetic and environmental influences, parts and wholes.

Somogyi, R., and Sniegolski, C. A. (1996). Modelling the complexity of genetic networks: Understanding multigenic and pleiotropic regulation. *Complexity* 1: 45–63.

URLs FOR RELEVANT SITES

Public initiatives to compile protein interaction into database for cell signaling pathways and network:

Cell Signaling Networks Database maintained by National Institute of Health Sciences (NIHS) (Japan). `http://geo.nihs.go.jp/csndb/`

The BioPathways Consortium. `http://www.biopathways.or/`

Tool for research in discrete dynamical networks:

Discrete Dynamics Lab. `http://www.santafe.edu/~wuensch/ddlab.html`

III Postgenomic Approaches

9 Predicting Protein Function and Networks on a Genomewide Scale

Edward M. Marcotte

WHAT IS PROTEIN FUNCTION?

Perhaps the most significant finding from the more than 80 genomes that have been sequenced as of 2002 has been the extent of our ignorance about the constituents of cells. In virtually every genome sequenced, the majority of genes have never been studied directly. In spite of this, for about half of the genes at least one near or distant relative has been studied, so we glean our knowledge from the activities of these relatives. Until recently such methods for extending information to proteins with similar sequences or structures (homology-based methods) have been the only form of inference about protein function.

Homology-based annotation, with algorithms such as BLAST (Altschul et al., 1997; http://www.ncbi.nlm.nih.gov/BLAST), has been wildly successful in extending knowledge from the small set of experimentally characterized proteins to the tens of thousands of proteins found in genome sequencing projects. However, these methods perform as one might expect: they provide information only for proteins with very closely related functions. They reveal little about proteins that work together but typically have unrelated sequences or structures. Thus, the homology-based methods cannot be used to reconstruct metabolic or signaling pathways or other protein interaction networks. That such a bias exists shows that there are different aspects to protein function; methods that reveal one aspect do not necessarily reveal others.

The two most important aspects of protein function, defined in figure 9.1, will be referred to as the *molecular function* and the *cellular function* of proteins. The homology-based methods tend to find only the molec-

Molecular Function

Structural Role

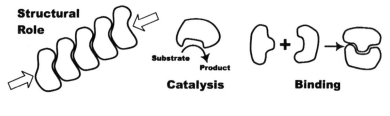

Catalysis

Substrate

Product

Binding

Cellular Function

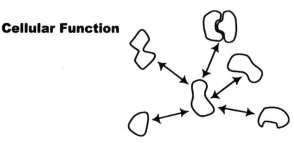

Figure 9.1 Two important components of protein function are the molecular (biochemical) function and the cellular (contextual) function (e.g., see Kim, 2000; Eisenberg et al., 2000). The molecular function of a protein is essentially the traditional view. It is the specific action that the protein engages in, such as binding, activation, inhibition, catalysis, fulfilling a structural role, etc. The cellular function is the system of interactions that the protein participates in, the context within which it operates. Other aspects of function include the intracellular location of proteins and the times and conditions under which proteins are expressed.

ular functions of proteins, but tell little about the context in which proteins operate. In fact, the context is crucial: proteins virtually never function alone in cells, but often interact with many partners. It has been estimated that an average protein will physically interact with 2–10 partners (Marcotte et al., 1999a). It can also be estimated that a protein will functionally interact—that is, participate in the same pathway—with even more proteins, perhaps two to three times the number of physical interactions. This interconnectedness is an important feature of the cellular organization and regulation of proteins. For this reason, protein networks are the subject of widespread study.

A new class of computational methods has been developed that finds the cellular function of proteins. This type of method is not based on comparisons of sequence or structure, but instead analyzes other

Edward M. Marcotte

attributes associated with genes. Broadly speaking, these nonhomology methods draw inferences about relationships between genes by analyzing the context in which the genes are found. This chapter will present an overview of these methods, along with a discussion of their applications for finding protein function, reconstructing cellular pathways, revealing new metabolic systems, and even revealing physical properties of proteins, such as their locations in cells.

GENOMES CONTAIN CONSIDERABLE INFORMATION ABOUT PROTEIN FUNCTION

It is easy to think only of the coding potential of genes, since that seems most immediately important for producing a protein. However, genes have many different properties besides their coding potential, and information about the relationships between genes is often encoded in these other properties. Important contextual properties of genes include their position and order on the chromosome, the flanking control regions, the distribution of homologues in other species, the occurrence of fusions between genes, and so on. Table 9.1 summarizes many such genomic sources of functional data and lists data derived from measurements of protein and mRNA expression patterns.

Just as homology-based methods analyze conserved sequences or structures to find proteins with related molecular function, so nonhomology methods analyze conserved contextual properties to find proteins with related cellular function. At the heart of nonhomology methods is the fact that proteins working together in the cell have shared constraints—they must be encoded by the same genome, they often are coregulated, they occasionally are fused into a single gene, they must at some point be coexpressed, and so on. Nonhomology methods exploit these constraints to identify proteins working together.

DISCOVERING PROTEIN FUNCTION FROM GENOMIC DATA

Finding Function from Domain Fusions

One of the most straightforward nonhomology methods needs large numbers of protein sequences but does not require complete genomes. It has been known for years that proteins encoded as separate genes in one organism often are found in another organism fused into a single

Table 9.1 Analysis of "contextual" information associated with genes

	Contextual Information	Applications
Information in genomes about the relationships between genes		
Information derived from a single genome	Intergenic distance	Operon reconstruction
	Intragenomic conservation of regulatory sequences	Operon and regulon reconstruction
Information derived from comparisons of multiple genomes	Distribution of sequence homologues among different organisms	Calculation of phylogenetic profiles for pathway reconstruction and cellular localization
	Conservation of relative gene position	Operon reconstruction
	Domain fusions	Pathway reconstruction
	Intergenomic conservation of regulatory regions	Identification of coregulated genes
Information in expression data about the relationships between genes		
Clustering genes by their expression profiles	mRNA expression profiles	Identification of coregulated genes and pathway or operon reconstruction
	Spatial expression profiles	Pathway reconstruction
	Protein expression profiles	Pathway reconstruction
Clustering genes by the expression levels of all other genes in one or more experiments	Genomewide expression as a gene phenotype	Pathway reconstruction

Beyond simply coding for genes and their regulatory sequences, genomes are rich in information about the relationships between genes. Analysis of this information allows reconstruction of cellular systems, pathways, and genetic networks. For the last entry, the expression of all other genes is used as the phenotype when the gene in question is disrupted. Genes are then clustered to maximally match their phenotypes.

Edward M. Marcotte

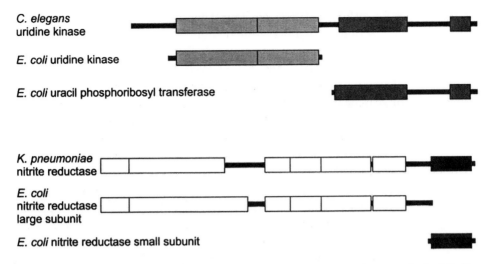

Figure 9.2 Two examples of the domain fusion or Rosetta Stone method of finding functional links. In each example the two lower proteins can be inferred to be functionally linked because of the existence of the top fusion protein. For example, if we did not already know that the *E. coli* nitrite reductase large and small subunits formed a hetero-complex, they could be inferred to be functionally linked after finding the *K. pneumoniae* fusion protein.

polypeptide. Two such examples are shown in figure 9.2. In each of the two examples, the separately encoded *E. coli* proteins are drawn beneath the fusion protein from another organism. In both cases, the *E. coli* proteins are members of the same pathway. In the bottom example, the nitrite reductase proteins physically interact to form an active nitrite reductase enzyme.

In fact, this trend is surprisingly common (Marcotte et al., 1999a; Enright et al., 1999), especially among metabolic proteins (Tsoka and Ouzounis, 2000). Thousands of such fusion events can be found—in yeast, more than 45,000 pairs of proteins can be found as fusion proteins in other organisms (Marcotte et al., 1999a). Almost universally, the cellular functions of the component proteins are very closely related. Searching systematically for these fusion events therefore rapidly generates functional links between proteins. For this reason, the fusion proteins have been called "Rosetta Stone" proteins for their ability to decode the functional links between component proteins (Marcotte et al., 1999a).

Rosetta Stone links are found by aligning a query protein's amino acid sequence against protein sequences from genomes or a large sequence database such as GenBank. The statistically significant hits from this search include sequence homologues and candidate Rosetta Stone proteins. These hits are then used as the query proteins for a second set of searches against the sequence database. The statistically significant hits from this second round of searches are then tested for similarity to the original query protein. Those second-round hits without sequence similarity to the original query protein are proteins with Rosetta Stone links to the original query protein.

Not all fusions convey the same degree of confidence in the resulting functional linkage. Individual domains have different propensities to participate in these gene fusion events, and many cell signaling domains, such as SH3 or tyrosine kinase domains, can be found fused into literally hundreds of different genes. These *promiscuous domains* still can be used to generate functional linkages, but it has been found that limiting the Rosetta Stone analysis to nonpromiscuous domains increases the functional similarity of the linked proteins. This filtering step can be performed either by explicitly forbidding links generated by promiscuous domains (Marcotte et al., 1999a; `http://www.doe-mbi.ucla.edu`) or by requiring strong sequence homology or even orthology between the individual proteins and the Rosetta Stone protein (Enright et al., 1999; Enright and Ouzounis, 2000). Regardless, it is possible to generate thousands of significant links between pairs of proteins in a genome by this method.

Finding Function from Coinheritance

One important consequence of the genomic revolution is the finding that genomes have mosaic compositions, containing genes with widely varying phylogenetic origins. This trend is especially strong among prokaryotes due to processes such as horizontal gene transfer (Jain et al., 1999; Koonin and Galperin, 1997), but is true to a considerable extent in eukaryotes as well (Marcotte et al., 2000). These variable phylogenetic origins of genes are another aspect of gene context for use in these analyses.

This phylogenetic diversity can be explicitly described for each gene by calculating its *phylogenetic profile* (Pellegrini et al., 1999, with related concepts in Gaasterland and Ragan, 1998; Huynen et al., 1998; Ouzounis and Kyrpides, 1996; and Tatusov et al., 2001). A phylogenetic profile describes the presence or absence of a gene across a set of organisms with sequenced genomes. Genes with similar phylogenetic profiles are therefore always inherited together or absent from the same organisms. This similarity is unlikely to happen by chance if enough species are examined, and such proteins are thus extremely likely to function together. For 30 genomes, there are about 2^{30}, or 10^9, possible phylogenetic profiles, making random matches of profiles unlikely. When enough different species are examined to be statistically significant, genes with similar phylogenetic profiles are inferred to be functionally linked.

Constructing a phylogenetic profile for a gene requires performing sequence alignments between that gene and all genes from each of the fully sequenced genomes. Because thousands of sequence alignments must be calculated, rapid alignment algorithms like BLAST (Altschul et al., 1997) are typically used. The phylogenetic profile of a gene is then calculated as a vector in which each entry represents a measure of sequence similarity between that gene and the most similar sequence match in a given genome. This measure of sequence similarity $S_{i,j}$ can be as simple as a binary code: $S_{i,j} = 1$ if a sequence homologue of gene i is present in genome j and $S_{i,j} = 0$ if no homologue exists. Alternatively, the measure of sequence similarity can be a real, valued measurement reflecting the degree of sequence similarity present. One such measure that has empirically been shown to work satisfactorily is $S_{i,j} = -1/\log(E)$, where E represents the expectation value from the sequence alignment between gene i and the top-scoring sequence match in genome j (Marcotte, 2000). Real-valued phylogenetic profiles calculated in this fashion are shown in figure 9.3.

Once a phylogenetic profile is calculated for each of the genes in a genome, functional links can then be inferred between genes with similar phylogenetic profiles. The simplest approach is to treat phylogenetic profiles as coordinate vectors positioning genes in a high-dimensional space, then calculating distances between genes, using such distance metrics as the Manhattan, Euclidean, or Mahalonobis distance. Genes positioned close together in space can be inferred to be coinherited, and therefore functionally linked. Another approach is to apply a statistical test such as a Fisher exact test on the binary phylogenetic profile vectors to identify coinherited genes.

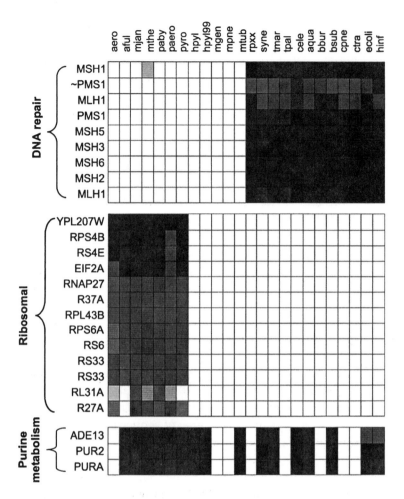

Figure 9.3 Examples of phylogenetic profiles for a number of yeast proteins. Each profile, drawn horizontally, indicates the degree of sequence similarity of a protein—for example, MLH1—to the most similar protein in each of the fully sequenced genomes (listed as abbreviations across the top.) Where there is no sequence homologue, the profile has a white square, and where there is a statistically significant sequence homologue, the square is colored to indicate the degree of homology, with black being most similar. Three functional classes of proteins are profiled; profiles are shared within a functional class but are quite distinct between classes.

As an alternative to calculating pairwise links, the genes can simply be clustered into coinheritance groups on the basis of similarity between their phylogenetic profiles. Many such clustering approaches have been developed in computer science and statistics for clustering points in high-dimensional spaces, such as *k*-means clustering, which are appropriate for this task.

Finding Function from Relative Gene Position

Another powerful method for finding functionally linked genes comes from examining the conservation of relative positions of genes in genomes (Dandekar et al., 1998; Tamames et al., 1997; Overbeek et al., 1999). As with the two previous methods, this aspect of gene context can be analyzed in a straightforward fashion. The essence of the method is that the order of genes in genomes tends to randomize over time. Therefore, if two genes have similar positions relative to one another in several genomes, the genes are likely to be functionally linked. In the simplest case, this means that the genes are immediate neighbors in several genomes, but the method could theoretically be extended to any separation between the genes. On-line tools for investigating the genomic neighbors of a gene include the Entrez genome (`http://www.ncbi.nlm.nih.gov/entrez/query.fcgi?db=Genome`) and WIT (`http://wit.mcs.anl.gov/WIT2`) databases.

This method exploits the trend for prokaryotic genes to be organized into operons, in which genes with a related function are clustered close together on the genome to allow coordinate transcription and translation of the genes. Operons seem to be uncommon in most eukaryotes, occurring mainly in unusual gene families such as the cadherins (Wu and Maniatis, 1999). However, in prokaryotes, operons are virtually the norm, and where genes from an operon are conserved in multiple species, this method allows very reliable functional links to be inferred. In fact, it has been shown that the observation of two genes as immediate neighbors in two reasonably unrelated organisms is sufficiently statistically significant to infer a functional link between the proteins (Overbeek et al., 1999), as diagrammed in figure 9.4A and B.

A rough calculation of significance goes as follows. Given two adjacent genes in a genome, we would expect by random chance to find the genes adjacent in a second genome of *n* genes, with all genes, but not gene order, conserved between the two genomes, only two times out of

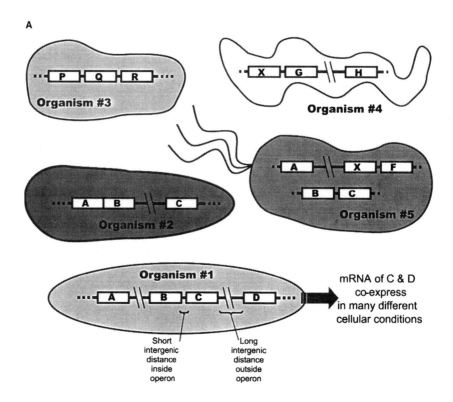

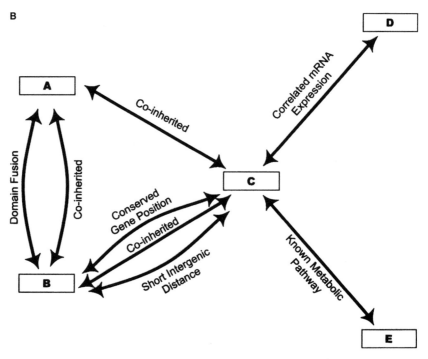

$n - 1$. So, given a typical bacterial genome of $n = 4000$ genes, we would expect to find the two genes adjacent with a random probability $p = 2/(3999)$, or 5×10^{-4}. Because of the ubiquity of operons and the low random likelihood of conserved neighbors, the coverage of this method can be quite high, and thousands of pairwise functional links can be generated (Huynen et al., 2000a).

Finding Function from Intergenic Distances

A second promising method has been described that analyzes gene position to find functional links between proteins. This method is explicitly formulated to detect operons and works by analyzing the number of nucleotides separating neighboring genes (Salgado et al., 2000). Genes organized in bacterial operons are cotranscribed on a single mRNA and translated in a coordinated fashion. This coordination of transcription and translation for the genes in an operon probably places a selective pressure on keeping genes close together, as compared to an absence of selection for adjacent genes not in the same operon. Thus, adjacent genes with short intergenic distances tend to be in the same operon; adjacent genes with long intergenic distances tend not to be.

One advantage of this method is that it can be performed for genes unique to a genome—no gene conservation is required for the method to operate. This ability to work on ORFans, the genes found only in a

Figure 9.4 (A) Comparing the genome of organism #1 with the genomes of several other organisms allows a number of inferences to be drawn about the relationships between the genes of organism #1. In the figure, genes, depicted as labeled white boxes, are arranged on the genomes, drawn as heavy black horizontal lines. First, the genes A and B can be found fused in organism #2, suggesting that A and B are functionally linked. Second, the genes A, B, and C are found in the same set of organisms (#1, 2, and 5) and are absent from the same set of organisms (#3 and 4). This coinheritance suggests A, B, and C are functionally linked. Third, genes B and C are neighbors in more than one genome, suggesting a selective pressure to maintain their relative positions. Likewise, the intergenic distance between genes B and C is much smaller than the typical intergenic distance, suggesting that B and C may belong to an operon. Fifth, the mRNA of genes C and D are coexpressed in many different experiments, suggesting C and D are coregulated or function together. Each of these inferences can be conceptualized as generating a functional linkage between two proteins. (B) The resultant network of functional links. Predicted networks can be compared can complemented by experimental networks, such as the experimentally derived link between C and E.

single genome (Fischer and Eisenberg, 1999), sets this method apart from the other genomic methods described above. In this respect, the analysis of intergenic distance has more in common with analysis of expression data, which can also provide functional links for ORFans.

Finding Function from Regulatory Regions

The last obvious contextual property of genes useful for assigning protein function is the presence of regulatory regions found outside of gene coding regions. These regulatory sites in DNA are recognized and bound by transcription factors, enhancers, and repressors to control the transcription of the neighboring genes. Because genes with related functions are often coregulated, it would seem reasonable to create functional links between genes with similar regulatory regions.

Unfortunately, regulatory regions are notoriously difficult to identify. Judging the similarity between them is equally difficult. Although consensus sequences have been identified for most major regulatory sites (e.g., see the Eukaryotic Promoter Database, `http://www.epd.isb-sib.ch`), the sites recognized by a given transcription factor or polymerase are often quite varied. Nonetheless, progress has been made in identifying shared regulatory regions upstream of coexpressed genes (Roth et al., 1998) and upstream of coinherited genes, Rosetta Stone linked genes, and genes coconserved in operons (Manson McGuire and Church, 2000). Since genes in these categories are often functionally related, it seems likely that the inverse process, clustering genes by their regulatory regions, will also yield functional information. An attempt at this process (Pavlidis et al., 2001) shows that such functional information is available, although the method is not currently as powerful as methods exploiting other sorts of contextual information. However, such analyses of regulatory regions are likely to improve dramatically with the explicit knowledge of transcription factor binding sites generated from DNA microarray mapping of transcription factor specificities (Iyer et al., 2001).

DISCOVERING PROTEIN FUNCTION FROM EXPRESSION DATA

The genomic analyses discussed above examine static genomes and draw inferences from the state of the genomes at one point in time. However, genomes and cells are dynamic systems, and considerable information can be gleaned about cellular systems by analyzing these dynamics. We might argue that genomes are dynamic on two time scales: the evolutionary and the immediate. The methods discussed above analyze events on the evolutionary time scale. Now we turn to events on the more immediate time scale.

Finding Function from mRNA Expression Patterns

DNA microarrays and EST sequencing have produced literally millions of discrete measurements of gene expression. This flood of data has in turn stimulated many analyses of gene expression profiles. In general, the analyses share the following form: A set of measurements of the expression of a number of genes under different conditions is available, from DNA microarrays (e.g., as in Lashkari et al., 1997), serial analysis of gene expression (SAGE; Velculescu et al., 1995), or expressed sequence tags (EST; Adams et al., 1991). Expression vectors are then constructed for the genes, each vector describing the expression of a given gene under a range of cellular conditions, cell types, genetic backgrounds, and so on. These expression vectors are then clustered to find genes with similar expression patterns (Eisen et al., 1998). Given enough independent experiments (>100) with sufficient variation in the conditions, genes clustered in this fashion tend to be functionally related (Marcotte et al., 1999b). Fortunately, unlike complete genome sequences, data of this sort are readily generated. It is possible to perform large numbers of microarray experiments, producing enough expression data to find statistically significant functional links. Many expression data sets are publicly available from sites such as the Stanford Microarray Database (`http://genome-www4.stanford.edu/MicroArray/SMD`).

A variation of this approach involves analysis of SAGE or EST libraries collected from various tissues and cell conditions (e.g., the dbEST database: `http://www.ncbi.nlm.nih.gov/dbEST/index.html`). In this approach, mRNAs from cells are reverse transcribed into cDNAs and sequenced. Since many thousands of mRNAs are typically sequenced, the EST or SAGE library is a fairly representative selection of the mRNAs present under those cellular conditions, and thus can serve in a fashion analogous to microarray expression measurements. EST and SAGE libraries vary widely in size and completeness, so calculations of expression vectors with their data are not entirely straightforward. However, related analysis can be performed, such as Guilt-by-Association, which essentially creates functional links between genes based on their copresence and coabsence from EST libraries (Walker et al., 1999).

Coexpression analyses have advantages and disadvantages in regard to genomic data for functional predictions. The primary disadvantage, beyond having to collect additional data, is that the functional inferences from coexpression are relatively weak until a large body of expression data is collected (Marcotte et al., 1999b). However, this is more than compensated for by the advantage of learning information about any gene for which expression can be detected, regardless of its conservation in other species. Coexpression analysis and the prediction of operons using intergenic distances (Salgado et al., 2000) are currently the only two computational methods capable of generating functional information for ORFans (Fischer and Eisenberg, 1999).

Finding Function from Spatial Expression Profiles

The expression methods discussed above typically give no information about the intracellular location of the expressed molecules. However, spatial expression data should be useful for pathway reconstruction, since we expect functionally linked proteins to be found at similar subcellular locations. Therefore, the converse will often be true: proteins that are always expressed at the same locations probably function together. This approach to finding protein function is quite technically demanding, but in spite of the difficulty, one group has collected such spatial expression data for more than 1750 genes expressed in *Xenopus* oocytes (Gawantka et al., 1998). More recently, the data have been incorporated into a database and methods to measure similarity between mRNA spatial expression patterns have been developed (Pollet et al., 2001). Although considerable work remains, this work establishes the viability of this method for generating functional information.

Finding Function from Protein Expression Profiles

Gathering expression data for an entire proteome, or all of the proteins encoded by a genome, is only now becoming feasible, due largely to the development of high-throughput mass spectrometric analyses of proteins (Shevchenko et al., 1996; Hunt et al., 1986; Gygi et al., 1999; Jensen et al., 2000). Although such protein expression data are not yet widely available, they will be a valuable complement to the mRNA expression data from EST libraries and chips. What is not yet clear is how well protein expression data will correlate with mRNA expression data.

Early results comparing protein expression by mass spectrometry and mRNA expression by SAGE suggested that mRNA and protein expression patterns are quite different (Gygi et al., 1999). Recent developments with DNA chips have allowed quantification of mRNAs being actively translated through analysis of polysomal mRNA fractions. These chip-based measurements of protein expression show strong correlation with chip-based measurements of mRNA expression (Joe DeRisi, personal communication). Nonetheless, it is likely that the protein expression patterns will hold considerable value for inferring protein function.

In theory, protein expression data can be analyzed similarly to mRNA expression data. It is likely that expression data will be collected for many of the proteins in a proteome over many different cellular conditions. As with mRNA expression data, these protein expression data will compose expression vectors that can be clustered and analyzed much as the mRNA data are.

However, protein expression data may contain an additional element absent from mRNA expression data: mass-spectrometric methods have the capability not only to measure protein expression levels but also to identify protein modifications. Posttranslational modifications of proteins are widespread in cells, both spontaneous unregulated events such as oxidation, and enzymatic modifications such as lipidation, phosphorylation, and ADP ribosylation. Such modifications often modify the activity or localization of the proteins. Thus, it seems likely that protein expression profiles will catalog not only expression patterns but also protein states, such as on, off, activated, repressed, and so on. These protein state vectors will provide a rich source of data for protein function prediction.

MEASURING PROTEIN FUNCTION AND TESTING PREDICTIONS

Before testing any of these predictive methods, one must develop a *metric* for measuring protein function. At first glance, protein function would seem difficult to quantify. However, several metrics have been developed that perform quite well, allowing optimization and calibration of the methods.

Perhaps the most obvious metric is that of testing that the methods recover known functional relationships. Using a database of known pathways, such as the KEGG (Kanehisa and Goto, 2000; `http://www.genome.ad.jp/kegg/kegg2.html`) or EcoCyc (Karp et al., 2000; `http://ecocyc.pangeasystems.com/ecocyc`) database of metabolic pathways, or the DIP database of protein interactions (Xenarios et al., 2001; `http://dip.doe-mbi.ucla.edu`), each method is evaluated by its coverage, the fraction of experimental links correctly predicted by the algorithm, and by its accuracy, the fraction of predicted links that are verified by an experimental link.

Unfortunately, the measurement of accuracy cannot be very exact, since our knowledge of experimental pathways is limited and few pathways are known completely. Thus, absence of a link from the experimental database does not necessarily mean the link is wrong. Due to this limited knowledge, we can measure false negative predictions accurately (failure to predict an experimental link), but cannot evaluate false positive predictions (prediction of a functional link where none exists). To some extent, the accuracy measurement, while not correct in an absolute sense, can be treated as a relative value for optimization and for comparisons between algorithms.

A second metric that performs well in practice is that of *key word* recovery or category matching (Marcotte et al., 1999b). For this approach, genes of known function are first classified into a limited set of functional categories. Many databases have such categorizations incorporated, sometimes explicitly (as in the MIPS database of yeast proteins; Mewes et al., 1998) and sometimes implicitly (as in the key words associated with proteins in the SWISSPROT protein sequence database; Bairoch and Apweiler, 2000). Testing predictions is then reduced to checking for agreement between the predicted and known key words or categories for each characterized protein, and finding the average agreement over all characterized proteins. An example is calculating

$$\langle \text{key word recovery} \rangle = \frac{1}{A} \sum_{i=1}^{A} \sum_{j=1}^{x} \frac{n_j}{N},$$

where x is the number of key words known for the protein i being

Edward M. Marcotte

tested, N is the number of key words predicted for the protein, and n_j is the number of times key word j from the protein's known annotation appears in the predicted key word list. The average key word recovery is calculated for all A characterized proteins.

For many of the predictive methods, the prediction is not of a given functional category but of a link between two proteins. In these cases, for all predicted protein pairs involving proteins of known function, the overlap between the key words or categories of the two proteins can be calculated with a function such as the Jaccard coefficient:

$$\langle \text{key word overlap} \rangle = \frac{1}{P} \sum_{i=1}^{P} \left(\frac{k_1 \cap k_2}{k_1 \cup k_2} \right),$$

where for each of the P pairs of linked, characterized proteins, the k_1 key words of one protein are compared against the k_2 key words of the linked protein partner. The number of key words in the intersection is divided by the number of key words in the union to give a normalized measure of the overlap between the two sets of key words. The value of this overlap averaged over all P pairs gives a measure of the accuracy of the prediction algorithm. To optimize and compare prediction algorithms, this measurement of method accuracy can be combined with the measured coverage of known pathways.

ASSIGNING PROTEINS TO FUNCTIONAL CATEGORIES

One simple way to implement these methods is to test if proteins belong to given functional categories (Pavlidis et al., 2001; Marcotte et al., 2000). To do this, an algorithm is trained to recognize the characteristics of proteins in a given functional category. Such a discrimination algorithm requires a set of quantitative features for each protein. Effectively, these features are treated as coordinates mapping the protein into a high-dimensional feature space. When the features are chosen appropriately, proteins belonging to a given functional category fall in a distinct region of this feature space and proteins from other functional categories fall in other regions.

Many of the contextual properties of genes can be interpreted as features. For example, the phylogenetic profile of a protein is a vector in which each element describes the degree of similarity of the protein to the most similar sequence in a given genome. When interpreted as a list of features, the phylogenetic profile describes the mapping of a protein into a *phylogenetic space*. The attributes of this space are the following: It is an n-dimensional space, where n is the number of genomes used to calculate the phylogenetic profile. The axes of the space are not orthogonal—some genomes are quite similar to each other, so some axes are more correlated than others. (If we choose, we can orthogonalize the space—for example, by applying a whitening transformation.) Last, proteins are not evenly distributed in this space. Certain systematic biases occur in the types of proteins encoded by a genome, and these in turn introduce biases in the genes' locations in phylogenetic space. For example, each genome contains a fraction of genes unique to that species; these genes all map to the same region of phylogenetic space. Likewise, certain genes are broadly conserved among only eukaryotes or prokaryotes—again, these genes all map to the same general region of phylogenetic space. However, proteins with a related function cluster in this space, as do eukaryotic proteins localized to similar cellular compartments (Marcotte et al., 2000).

A discrimination algorithm defines a set of boundaries in this high-dimensional space that separate proteins with the desired function from all other proteins. Numerous algorithms have been adapted from statistics and computer science for this purpose, including Bayesian classifiers (elegantly described in Mosteller and Wallace, 1984), support vector machines (Pavlidis et al., 2001), neural network discriminators, and linear discrimination functions (Marcotte et al., 2000). The advantage of this method of predicting function is that one can test for very specific functions, as well as calculate the degree of confidence in the results.

INTEGRATING METHODS TO DISCOVER PROTEIN FUNCTIONAL AND INTERACTION NETWORKS

The discrimination algorithms described above work under the assumption that a set of functionally related proteins is known, and more proteins with the same function are desired. In this approach, the algorithms must be trained on a set of positive examples, proteins whose functions are known to match the desired function, as well as on a set of negative examples.

However, a naive approach can be useful to discover what trends are in the data and to look for naturally occurring clusters. The naive

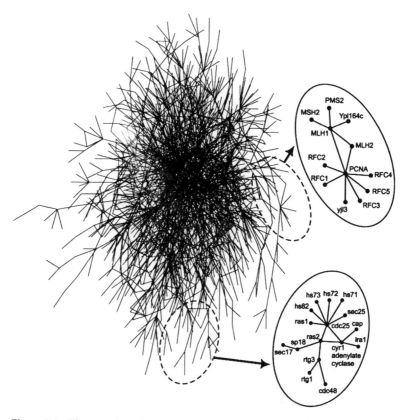

Figure 9.5 The proteins of yeast interact in an extensive network. Here, the vertices of this graph are 1722 yeast proteins participating in 2612 experimentally observed interactions, drawn as edges connecting the interacting partners. Two regions are expanded to show an interaction network involving the ras protein and an interaction network involving several DNA replication factors (RFC1–5). Many experimental techniques are represented, including high-throughput two-hybrid interaction screens (Uetz et al., 2000; Ito et al., 2000), mass spectrometry, and co-immunoprecipitation. The interactions are available courtesy of Ioannis Xenarios, curator of the Database of Interacting Proteins (Xenarios et al., 2001).

approach is also biologically motivated: it is now becoming apparent that proteins are organized into large interaction networks in the cell. One such experimentally derived protein interaction network is shown for the proteins of yeast in figure 9.5, derived from high-throughput measurements of protein interactions (Uetz et al., 2000; Ito et al., 2000) and from mining biological literature for all previously known yeast protein interactions (Xenarios et al., 2001). Such networks reinforce the notion that proteins never work alone. Ideally, the predictive methods should reveal exactly these sorts of networks.

To discover such networks, the predictive methods can be applied to produce functional links between pairs of proteins. Although the links are generated in a pairwise fashion, extensive networks of proteins result when links are calculated for all of the genes in a genome. Different types of networks are calculated, depending on the method used. For example, mRNA expression links may produce coexpression networks, and phylogenetic profiles will produce coinheritance networks. However, networks provide a logical framework for combining methods. Using the metrics described earlier, each method can be optimized to link proteins with a comparable degree of confidence. Then, links generated by each method can be combined to create a functional interaction network. A simplified example is diagrammed in figure 9.4B, derived from the gene context information for one the organisms (#1) in figure 9.4A.

Actual networks calculated for all of the proteins encoded in a genome are much more complicated. Figure 9.6 shows such a predicted functional network for 2240 proteins of yeast. Inspection of the network shows that it has considerable diversity in its structure, with many highly connected subnetworks. Examination of such subnetworks shows reasonable correspondence to many known pathways (e.g., see Pellegrini et al., 1998; Marcotte, Pellegrini, Ng, et al., 1999; and Marcotte, Pellegrini, Thompson, et al., 1999). Uncharacterized proteins can therefore be assigned function by linking them with known pathways. This approach allowed preliminary assignment of functions to more than half the uncharacterized proteins of yeast (Marcotte, Pellegrini, Thompson, et al., 1999; http://www.doe-mbi.ucla.edu/) and to 10% of the genes of *M. genitalium* (Huynen et al., 2000b).

Analysis of these predictive networks and their correspondence to metabolic, signaling, and interaction networks is an ongoing area of study. Open topics of study include defining subnetworks, cliques, and network properties; determining which functional links correspond to physical interactions and which have other interpretations; and dynamic models of networks. Predictive networks can be incorporated into metabolic pathway models, such as those discussed in chapter 10 or those incorporated into the E-cell project, described in chapter 11.

Figure 9.6 A network of predicted functional links between yeast proteins. As in figure 9.5, proteins are drawn as vertices of the graph, and functional links are drawn as edges between functionally linked proteins. In all, 2240 proteins are shown participating in 12,012 functional links, as calculated from phylogenetic profiles (adapted from Marcotte, 2000).

For these models, predictive networks may be especially useful for completing input pathways known only partially from experiment.

DISCOVERING NEW METABOLIC SYSTEMS

An especially tantalizing aspect of the study of protein networks is the discovery of novel cellular systems. Molecular biology has until recently generated knowledge about proteins one at a time, each researcher studying the system of his or her desire. The overall effect has been a somewhat random patterning of knowledge over the proteome.

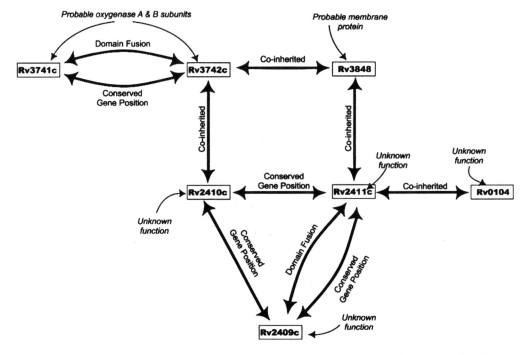

Figure 9.7 Novel pathways are revealed in computationally predicted networks. Shown here is a network of *M. tuberculosis* genes linked together by a combination of predictive methods. Multiple methods support each other in linking the genes, increasing confidence that the proteins participate in the same pathway. At the time of this writing, functions were unknown for all of the genes, with the exception of homology of Rv3741c and Rv3742c to oxygenase subunits. This homology suggests that the genes in the network may participate in a novel metabolic pathway in *M. tuberculosis*.

However, the functional and physical interaction networks give the best and most complete estimates of cellular pathways and systems. Examining the networks shows exactly which systems have been well studied and which have been neglected entirely. By searching for such unstudied systems, the network analysis allows systematic discovery of novel pathways.

One such novel system from *Mycobacterium tuberculosis* is diagrammed in figure 9.7. This system was found in a search for tightly functionally linked but unannotated genes. The genes are linked by a number of redundant functional linkages, increasing the likelihood that the genes really function together. Of the seven genes in this putative pathway, none have a known function, although two are estimated from sequence homology to be subunits of an uncharacterized oxygenase. Here, the network analysis reveals only the cellular functions, but not the molecular functions, of the proteins. We can only speculate, based upon the oxygenase proteins, that this system is a novel metabolic pathway in *M. tuberculosis*. Defining all such new systems is the first step; what follows is perhaps the harder work of characterizing and understanding the new systems.

REFERENCES

Adams, M. D., Kelley, J. M., Gocayne, J. D., Dubnick, M., Polymeropoulous, M. H., Xiao, H., Merril, C. R., Wu, A., Olde, B., Moreno, R. F., et al. (1991). Complementary DNA sequencing: Expressed sequence tags and human genome project. *Science* 252: 1651–1656.

Altschul, S. F., Madden, T. L., Schaffer, A. A., Zhang, J., Zhang, Z., Miller, W., and Lipman, D. J. (1997). Gapped BLAST and PSI-BLAST: A new generation of protein database search programs. *Nucleic Acids Res.* 25: 3389–3402.

Bairoch, A., and Apweiler, R. (2000). The SWISS-PROT protein sequence database and its supplement TrEMBL in 2000. *Nucleic Acids Res.* 28: 45–48.

Dandekar, T., Snel, B., Huynen, M., and Bork, P. (1998). Conservation of gene order: A fingerprint of genes that physically interact. *Trends Biochem. Sci.* 23(9): 324–328.

Eisen, M. B., Spellman, P. T., Brown, P. O., and Botstein, D. (1998). Cluster analysis and display of genome-wide expression patterns. *Proc. Natl. Acad. Sci. USA* 95: 14863–14868.

Eisenberg, D., Marcotte, E. M., Xenarios, I., and Yeates, T. O. (2000). Protein function in the post-genomic era. *Nature* 405: 823–826.

Enright, A. J., Iliopoulos, I., Kyrpides, N. C., and Ouzounis, C. A. (1999). Protein interaction maps for complete genomes based on gene fusion events. *Nature* 402: 86–90.

Enright, A. J., and Ouzounis, C. A. (2000). GeneRAGE: A robust algorithm for sequence clustering and domain detection. *Bioinformatics* 16: 451–457.

Fischer, D., and Eisenberg, D. (1999). Finding families for genomic ORFans. *Bioinformatics* 15: 759–762.

Gaasterland, T., and Ragan, M. A. (1998). Constructing multigenome views of whole microbial genomes. *Microb. Comp. Genomics* 3: 177–192.

Gawantka, V., Pollet, N., Delius, H., Vingron, M., Pfister, R., Nitsch, R., Blumenstock, C., and Niehrs, C. (1998). Gene expression screening in *Xenopus* identifies molecular pathways, predicts gene function and provides a global view of embryonic patterning. *Mech. Dev.* 77: 95–141.

Gygi, S. P., Rochon, Y., Franza, B. R., and Aebersold, R. (1999). Correlation between protein and mRNA abundance in yeast. *Mol. Cell. Biol.* 19: 1720–1730.

Hunt, D. F., Yates, J. R. III, Shabanowitz, J., Winston, S., and Hauer, C. R. (1986). Protein sequencing by tandem mass spectrometry. *Proc. Natl. Acad. Sci. USA* 83: 6233–6237.

Huynen, M., Dandekar, T., and Bork, P. (1998). Differential genome analysis applied to the species-specific features of *Helicobacter pylori*. *FEBS Lett.* 426: 1–5.

Huynen, M., Snel, B., Lathe, W. III, and Bork, P. (2000a). Exploitation of gene context. *Curr. Opinion Struct. Biol.* 10: 366–370.

Huynen, M., Snel, B., Lathe, W. III, and Bork, P. (2000b). Predicting protein function by genomic context: Quantitative evaluation and qualitative inferences. *Genome Res.* 10: 1204–1210.

Ito, T., Tashiro, K., Muta, S., Ozawa, R., Chiba, T., Nishizawa, M., Yamamoto, K., Kuhara, S., and Sakaki, Y. (2000). Toward a protein-protein interaction map of the budding yeast: A comprehensive system to examine two-hybrid interactions in all possible combinations between the yeast proteins. *Proc. Natl. Acad. Sci. USA* 97: 1143–1147.

Iyer, V. R., Horak, C. E., Scafe, C. S., Botstein, D., Snyder, M., and Brown, P. O. (2001). Genomic binding sites of the yeast cell-cycle transcription factors SBF and MBF. *Nature* 409: 533–538.

Jain, R., Rivera, M. C., and Lake, J. A. (1999). Horizontal gene transfer among genomes: The complexity hypothesis. *Proc. Natl. Acad. Sci. USA* 96: 3801–3806.

Jensen, P. K., Pasa-Tolic, L., Peden, K. K., Martinovic, S., Lipton, M. S., Anderson, G. A., Tolic, N., Wong, K. K., and Smith, R. D. (2000). Mass spectrometric detection for capillary isoelectric focusing separations of complex protein mixtures. *Electrophoresis* 21: 1372–1380.

Kanehisa, M., and Goto, S. (2000). KEGG: Kyoto Encyclopedia of Genes and Genomes. *Nucleic Acids Res.* 28: 27–30.

Karp, P. D., Riley, M., Saier, M., Paulsen, I. T., Paley, S. M., and Pellegrini-Toole, A. (2000). The EcoCyc and MetaCyc databases. *Nucleic Acids Res.* 28: 56–59.

Kim, S. H. (2000). Structural genomics of microbes: An objective. *Curr. Opinion Struct. Biol.* 10: 380–383.

Koonin, E. V., and Galperin, M. Y. (1997). Prokaryotic genomes: The emerging paradigm of genome-based microbiology. *Curr. Opinion Genet. Dev.* 7: 757–763.

Lashkari, D. A., De Risi, J. L., McCusker, J. H., Namath, A. F., Gentile, C., Hwang, S. Y., Brown, P. O., and Davis, R. W. (1997). Yeast microarrays for genome wide parallel genetic and gene expression analysis. *Proc. Natl. Acad. Sci. USA* 94: 13057–13062.

Manson McGuire, A., and Church, G. M. (2000). Predicting regulons and their cis-regulatory motifs by comparative genomics. *Nucleic Acids Res.* 28: 4523–4530.

Marcotte, E. M. (2000). Computational genetics: Finding protein function by non-homology methods. *Curr. Opinion Struct. Biol.* 10: 359–365.

Marcotte, E. M., Pellegrini, M., Ng, H. L., Rice, D. W., Yeates, T. O., and Eisenberg, D. (1999). Detecting protein function and protein-protein interactions from genome sequences. *Science* 285: 751–753.

Marcotte, E. M., Pellegrini, M., Thompson, M. J., Yeates, T. O., and Eisenberg, D. (1999). A combined algorithm for genome-wide prediction of protein function. *Nature* 402: 83–86.

Marcotte, E. M., Xenarios, I., van der Bliek, A. M., and Eisenberg, D. (2000). Localizing proteins in the cell from the phylogenetic profiles. *Proc. Natl. Acad. Sci. USA* 97: 12115–12120.

Mewes, H. W., Hani, J., Pfeiffer, F., and Frishman, D. (1998). MIPS: A database for protein sequences and complete genomes. *Nucleic Acids Res.* 26: 33–37.

Mosteller, F., and Wallace, D. L. (1984). *Applied Bayesian and Classical Inference: The Case of the Federalist Papers*. New York: Springer-Verlag.

Ouzounis, C., and Kyrpides, N. (1996). The emergence of major cellular processes in evolution. *FEBS Lett.* 426: 1–5.

Overbeek, R., Fonstein, M., D'Souza, M., Pusch, G. D., and Maltsev, N. (1999). The use of gene clusters to infer functional coupling. *Proc. Natl. Acad. Sci. USA* 96: 2896–2901.

Pavlidis, P., Furey, T. S., Liberto, M., Haussler, D., and Grundy, W. N. (2001). Promoter region-based classification of genes. *Proc. Pac. Symp. Biocomput.* 6: 151–163.

Pavlidis, P., Weston, J., Cai, J., and Grundy, W. N. (2001). Gene functional classification from heterogeneous data. *Proceedings of the Fifth International Conference on Computational Molecular Biology*. pp. 242–248.

Pellegrini, M., Marcotte, E. M., Thompson, M. J., Eisenberg, D., and Yeates, T. O. (1999). Assigning protein functions by comparative genome analysis: Protein phylogenetic profiles. *Proc. Natl. Acad. Sci. USA* 96: 4285–4288.

Pollet, N., Schmidt, H. A., Gawantka, V., Niehrs, C., and Vingron, M. (2000). *In silico* analysis of gene expression patterns during early development of *Xenopus laevis*. *Proc. Pac. Symp. Biocomput.* 5: 443–454.

Roth, F. P., Hughes, J. D., Estep, P. W., and Church, G. M. (1998). Finding DNA regulatory motifs within unaligned noncoding sequences clustered by whole-genome mRNA quantitation. *Nat. Biotechnol.* 16: 939–945.

Salgado, H., Moreno-Hagelsieb, G., Smith, T. F., and Collado-Vides, J. (2000). Operons in Escherichia coli: Genomic analyses and predictions. *Proc. Natl. Acad. Sci. USA* 97: 6652–6657.

Shevchenko, A., Jensen, O. N., Podtelejnikov, A. V., Sagliocco, F., Wilm, M., Vorm, O., Mortensen, P., Shevchenko, A., Boucherie, H., and Mann, M. (1996). Linking genome and proteome by mass spectrometry: Large-scale identification of yeast proteins from two-dimensional gels. *Proc. Natl. Acad. Sci. USA* 93: 14440–14445.

Tamames, J., Casari, G., Ouzounis, C., and Valencia, A. (1997). Conserved clusters of functionally related genes in two bacterial genomes. *J. Mol. Evol.* 44: 66–73.

Tatusov, R. L., Natale, D. A., Garkavtsev, I. V., Tatusova, T. A., Shankavaram, U. T., Rao, B. S., Kiryutin, B., Galperin, M. Y., Fedorova, N. D., and Koonin, E. V. (2001). The COG database: New developments in phylogenetic classification of proteins from complete genomes. *Nucleic Acids Res.* 29: 22–28.

Tsoka, S., and Ouzounis, C. A. (2000). Prediction of protein interactions: Metabolic enzymes are frequently involved in gene fusion. *Nat. Genetics.* 26: 141–142.

Uetz, P., Giot, L., Cagney, G., Mansfield, T. A., Judson, R. S., Knight, J. R., Lockshon, D., Narayan, V., Srinivasan, M., Pochart, P., et al. (2000). A comprehensive analysis of protein-protein interactions in *Saccharomyces cerevisiae. Nature* 403: 623–627.

Velculescu, V. E., Zhang, L., Vogelstein, B., and Kinzler, K. W. (1995). Serial analysis of gene expression. *Science* 270: 484–487.

Walker, M. G., Volkmuth, W., and Klingler, T. M. (1999). Pharmaceutical target discovery using Guilt-by-Association: Schizophrenia and Parkinson's disease genes. In *Proceedings of the 7th International Conference on Intelligent Systems for Molecular Biology.* MenloPark, Calif: AAAIPress, pp. 282–286.

Wu, Q., and Maniatis, T. (1999). A striking organization of a large family of human neural cadherin-like cell adhesion genes. *Cell* 97: 779–790.

Xenarios, I., Fernandez, E., Salwinski, L., Duan, X. J., Thompson, M. J., Marcotte, E. M., and Eisenberg, D. (2001). DIP: The database of interacting proteins. 2001 update. *Nucleic Acids. Res.* 29: 239–241.

SUGGESTED READING

Brent, R. (1999). Functional genomics: Learning to think about genomic expression data. *Curr. Biol.* 9: R338–R341.

Brown, P. O., and Botstein, D. (1999). Exploring the new world of the genome with DNA microarrays. *Nat. Genetics* 21: 33–37.

Cho, R. J., and Campbell, M. J. (2000). Transcription, genomes, function. *Trends in Genetics* 16: 409–415.

Legrain, P., Jestin, J.-L., and Schächter, V. (2000). From the analysis of protein complexes to proteome-wide linkage maps. *Curr. Opinion Biotechnol.* 11: 402–407.

Pandey, A., and Mann, M. (2000). Proteomics to study genes and genomes. *Nature* 405: 837–846.

URLs FOR RELEVANT SITES

The web-based Blast server, which allows searches for sequence similarity between proteins or DNA sequences, can be found at the NCBI web site:
`http://www.ncbi.nlm.nih.gov/BLAST`

Lists of promiscuous domains are available through the UCLA Structural Biology lab web site, as are functions predicted by non-homology methods for many of the genes of yeast:
`http://www.doe-mbi.ucla.edu`

Improved nonhomology predictions can be found with the Protein Link Explorer (PLEX) at the University of Texas bioinformatics web site:
`http://polaris.icmb.utexas.edu/main.html`

Tools for investigating the genomic neighborhood of a gene include the Entrez genome web site: `http://ww.ncbi.nlm.nih.gov/entrez/query.fcgi?db=Genome` and the WIT database: `http://wit.mcs.anl.gov/WIT2`

The Eukaryotic Promotor database lists the consensus regulatory sequences derived from promotors of many eukaryotic genes: `http://www.epd.isb-sib.ch`

DNA microarray protocols, numerous experiments, and data are available from the Stanford Microarray Database:
`http://genome-www4.stanford.edu/MicroArray/SMD`

Additional measurements of gene expression are available in Expressed Sequence Tag databases such as dbEST: `http://www.ncbi.nlm.nih.gov/dbEST/index.html`

Many known metabolic pathways and networks have been characterized in the KEGG database: `http://www.genome.ad.jp/kegg/kegg2.html`

EcoCyc database: `http://ecocyc.pangeasystems.com/ecocyc`

Database of Interacting Proteins: `http://dip.doe-mbi.ucla.edu`

Last, several sequence databases also provide functional annotation of the genes that can be used in benchmarking programs that predict gene function. Among these annotated databases are

MIPS: `http://www.mips.biochem.mpg.de/`

Swiss-Prot: `http://ca.expasy.org/sprot/`

Gene Ontology Consortium: `http://www.geneontology.org`

10 Metabolic Pathways

Steffen Schmidt and Thomas Dandekar

ENZYMATIC REACTIONS

Life is a process far from thermodynamic equilibrium. Different enzymatic chains and networks have to operate continuously to prevent chemical or thermodynamic equilibrium from occurring, for that would mean the death of the organism. The analysis of metabolic pathways tries to understand and investigate the enzyme chains involved in this process.

A qualitative focus considers which enzymes are involved, which pathways are possible in a complex network (a nontrivial question), and what happens if one enzyme is knocked out or a new one is added to the set. Techniques developed in our laboratory to analyze metabolic pathways qualitatively will be the focus of this chapter. It should be noted, however, that there is also a detailed quantitative perspective from which to analyze metabolic pathways in their conversions and transitions—but it needs much more experimental data on the considered enzymes. It includes metabolite concentrations, substrate and product affinities, and kinetic data on enzymes such as their turnover number.

Isotope labeling of metabolites and NMR are advanced tools to provide experimental data for such studies (Christensen and Nielsen, 2000). These questions are only briefly discussed in this chapter. The interested reader is referred to the excellent textbooks and reviews on these questions (e.g., Fell, 1997; Hatzimanikatis, 1999; Stefanopoulos, 1999). Dynamics of metabolites may also be tackled in comprehensive cellular simulations (e.g., Womble and Rownd, 1986). (This rapidly developing area is discussed in chapter 11.)

Another aspect of all enzymatic reactions is that they can be divided into two types. There are biochemical reactions, that can proceed from substrate to product and vice versa: these are reversible reactions. There are also reactions that proceed in only one direction: these are irreversible reactions. In most cases the latter is achieved as the product is either moved from the cell into outer compartments or by rapid conversion of the product further down the enzymatic chain. There are over 2000 known biochemical reactions. Good references are the ENZYME database (`http://www.expasy.ch/enzyme/index.html`) and the BRENDA database (`http://www.brenda.uni-koeln.de`).

If there is a pathway of enzymatic reactions, the flow through the system is called the metabolic flux (e.g., the rate of flow in micromoles per minute) of substrates converted into products.

Another very general division of enzymes is between energy-producing (exergonic reactions) and energy-consuming (endergonic reactions). Similarly, on the metabolic level, metabolic pathways can conveniently be divided into catabolic (metabolite degrading; any type of metabolic "digestion") and anabolic (producing more complex biological metabolites). Anabolic pathways usually contain many endergonic reactions because they require a great deal of energy to create the more complex metabolites. They are generally more extensive in their network and enzymatic reactions, and occur more often in all autotrophic organisms (such as plants, where the required energy is supplied by photosynthesis). Relatively few metabolites serve as starting materials for a variety of products.

In contrast, all animals (including man), as well as fungi, parasites, and saprophytic organisms, rely for their energy supply on catabolic pathways, which usually contain more exergonic enzymatic reactions. Few common intermediates are produced from a large number of substrates. An important first end product is often acetyl-CoA, and ultimately, via oxidative pathways, oxidative phosphorylation of all available reduction equivalents (NADH, FAD, etc.) to generate a maximum of ATP. Similarly, NADPH and ATP serve as major high-energy components in transfer reactions in between.

A metabolic pathway is thus a series of connected enzymatic reactions that produce a specific product. Ample examples are provided by the KEGG metabolic database (`http://www.genome.ad.jp/kegg/pathway/map/map01100.html`). Furthermore, pathways may be branched or interconnected. Delineating a pathway from the metabolic network has until now mainly been done using traditional or chemical knowledge. Algorithms are very helpful in gaining rationality and precision. The abundance of enzymes varies with the kind of organism, cell type, nutrient status, and developmental stage. Another aspect, which can be partly revealed by meticulous genome comparison and annotation, is the detection of isoenzymes: the same enzyme is present in different varieties, depending on tissue or developmental state; in prokaryotes, even in the same cell. Different regulatory domains often confer different regulation of isoenzymes, and thus tissue-specific or condition-specific metabolization for the same type of reaction can be achieved. An example is lactate dehydrogenase, in which heart-specific and muscle-specific varieties exist, allowing for optimal adaptation of the metabolism in the two different types of tissues.

Problems we will discuss are pathway identification and completeness, identification of alternative pathways and all possible fluxes through a given metabolic network, and better understanding of the importance and the specific biological significance of pathway variations.

Pathways are the next level of complexity after the protein reading frame. On the other hand, they are only an abstraction, because in reality there are always metabolic networks. In that respect, it is important to have tools for their analysis as well as their synthesis.

FINDING THE ROUTE THROUGH THE METABOLIC NETWORK

Defining Metabolic Pathways

Cellular reactions are highly connected and entangled (figure 10.1; Jeong et al., 2000). In order to get a good overview of the complexity of metabolism, one can use standard literature and databases available on the Internet (e.g., KEGG, EcoCyc, Boehringer Mannheim Map). However, metabolism is much more flexible than the well-known set of pathways available from biochemistry textbooks. Using known information about metabolism, computational calculations can help to find new possible routes in the biochemical network that can then be proven by experimental approaches such as isotope labeling of metabolites, use of metabolic inhibitors, and knocking out of enzymes. Through chemical reasoning or tradition, standard pathways such as glycolysis are known for well-known metabolites. However, given the network

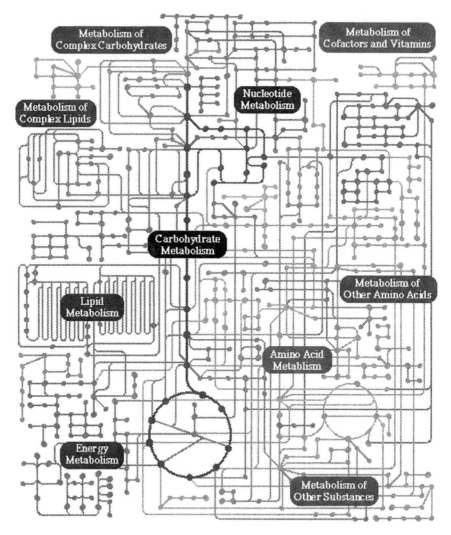

Figure 10.1 The complexity and interconnectedness of metabolic pathways is sketched in this simplified cellular pathway map, which considers only textbook pathways.

nature of enzymatic interactions, are these the only possible routes through an enzyme network? In a similar vein, new genome expression data try to predict coregulation of enzymes, and in this way provide a completely different approach to pathway definition. Can such coregulated enzymes be called a pathway? To decide this on a more rational basis, a more algorithmic definition of pathway is required.

S. Schmidt and T. Dandekar

Three different approaches are generally used to map the pathways in a biochemical network from scratch. They do not require information about enzyme kinetics, affinity, or metabolite concentrations—but this means each of them is only stating whether a certain metabolic route (a set of enzymes working together) is accessible to the set of enzymes present in the cell. It does not give information on the extent to which a certain metabolic path is used in the cell under a given metabolic condition.

A first way to identify a pathway is by synthesis. Transformation routes from a substrate to a product are found by successive addition of reactions, very similar to the way in which the major, well-known pathways were found by chemists. This can be done systematically, and details of this type of analysis are available in Seressiotis and Bailey (1986).

Elementary Mode Analysis

A powerful tool is the use of elementary flux modes (Schuster and Hilgetag, 1994). This will be examined now in more detail.

An elementary mode is the minimal set of enzymes that can operate at steady state, with the enzymes weighted by the relative flux they carry. "Minimal" is to be understood in the sense that if only the enzymes belonging to this set are operating, complete inhibition of a further enzyme of this set would lead to cessation of any steady-state flux in the system. With the help of the elementary modes, it can be determined whether a stoichiometrically balanced path exists between a particular set of substrates and products—no matter how complex and entangled a metabolic network may be. Actually, it is not sufficient to construct linear pathways just by following different metabolites, because by-products that cannot be excreted by the cell would then have to be balanced by additional reactions that produce further by-products, and so on. After this the full set of nondecomposable steady-state flows that the network can show, including cyclic flows (Leiser and Blum, 1987), is enumerated. An actual steady-state flux pattern such as measured by an experiment, will always be a non-negative linear combination of these modes. Thus the living cell is described as a mixture of pure states of these elementary modes that form a unique set.

The stoichiometry of the network is critical for identifying the elementary flux modes (Clarke, 1981), the nondecomposable modes that allow a steady-state flux from substrates to products. At first the enyzmes involved are abbreviated as is convenient, and from biochemical knowledge it is determined in which direction the reactions catalyzed would proceed. For example, in modeling a central part of the metabolism involving glycolysis and the pentose phosphate cycle, one could write

—ENZREV
PGI ALD TPI ... (for phosphoglucoseisomerase, aldolase, and triosephosphate isomerase, etc.)
—ENZIRREV
HK PFK ... (hexokinase, phosphofructokinase, etc., these are irreversible)

Next the external and internal metabolites are defined. For an external metabolite it is assumed that there is sufficient buffering capacity either in the cell or in the enviroment so that the steady-state condition is always fulfilled for this metabolite. In contrast, for any internal metabolite all reactions have to be properly represented by stoichiometric equations in the model because it has to be properly supplied in stable flux from the enzymes modeled in the system. For our example this would be

—METINT
E4P S7P ... (erythrose 4-phosphate, seduheptulose 7-phosphate, and so on are internal; the example metabolites are internally created by the pentose phosohate cycle)
—METEXT
CO_2 NADP NADPH ... (and so on; for example cofactors are involved in so many other cellular reactions that they may be considered to be externally created and buffered)

Finally, all enyzmatic reactions involved in the transformation are written down:

—CAT
HK: GLC + ATP = G6P + ADP
PGI: G6P = F6P
PFK: F6P + ATP = FDP + ADP
· · ·
(the enzymatic reactions of hexokinase, phosphoglucose isomerase, phosphofructokianse, and so forth)

The algorithm calculating the elementary modes from this information is now briefly sketched. (A detailed computational protocol is at `http://bms-mudshark.brookes.ac.uk/algorithm.pdf`; examples and application to complex networks are described in Schuster et al., 2000). The algorithm has been implemented as computer programs in Smalltalk (program EMPATH, John Woods, Oxford) and C (program METATOOL, Pfeiffer et al., 1999). Both are available from `ftp://bmshuxley.brookes.ac.uk/pub/mca/software/ibmpc`. The programs start from a list of reaction equations and a declaration of reversible and irreversible reactions, and of internal and external metabolites. As in several other metabolic simulators, this list is automatically translated into a stoichiometric matrix (for an explanation of this and related terms, see `http://www.biologie.hu-berlin.de/biophysics/Theory/tpfeiffer/metatool.html`). This matrix is then transposed and augmented with the identity matrix, to give a matrix called the initial tableau. From this, further tableaux are consecutively computed by pairwise linear combination of rows so that the columns of the transposed stoichiometric matrix successively become null vectors. This procedure corresponds to ensuring that the steady-state constraint is satisfied for each metabolite in turn.

Different metabolic pathways can be analyzed in this way. This assumes that there has already been a detailed stocktaking of the available enzymes. This will be explained by two methods, pathway alignment and pathway completion by genome annotation.

Apart from just enumerating paths, applications of elementary mode analysis include the following:

—Testing whether a set of enzymes allows production of a desired product

—Detecting nonredundant pathways and enzyme activities (important for drug design)

—Detecting pathways with maximal molar yields (important for biotechnology and metabolic engineering) as well as alternatives with nearly maximal yields

—Genome comparisons—it rapidly summarizes whether pathways are complete, and detects gaps and inconsistencies in the annotation

—Medical or pathophysiological assessment of the impact of enzyme deficiencies.

However, for several pharmacological applications, elementary mode analysis is quite interesting:

INPUT:

Enzymes and substrates from glycolysis and
pentose phosphate cycle.

OUTPUT:

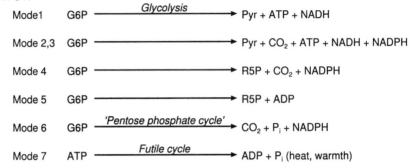

Figure 10.2 Sketch of the elementary modes calculated for the pentose phosphate cycle
and glycolysis. After program input (for details, see text), the algorithm METATOOL calculates seven elementary modes (listed below the input; products are given only qualitatively; modes 2 and 3 differ in their enzyme connectivity and their exact stoichiometry).
This illustrates that even such a simple network contains more modes of operation than
biochemical textbook knowledge would suggest.

—Identifying toxic metabolites

—Following pharmacologically active, intermediary metabolites and
products accruing or created by the enzyme machinery of the body
after, for example, painkilling drugs or sedatives are metabolized

In metabolic engineering, the method also allows *in silico* tests to determine whether artificially introduced pathways are likely to function
stoichiometrically. This could concern redox balance.

Furthermore, an improved analysis of the effects of adding other
substrates to the network is interesting. In some medically quite interesting cases this may lead to either potentiation or reduction of the
effect of a drug.

Textbook descriptions of standard pathways can now be made much
more precise or complete by results from elementary mode analysis
(figure 10.2). Thus, analyzing the system composed of glycolysis and
the pentose phosphate cycle in some more detail (Schuster et al., 2000),
we could show that a total of seven modes can be obtained. Besides the
textbook pathways (modes 3 to 6), some additional modes of conversion can be found: Mode 1 represents the usual glycolytic pathway.

Modes 2 and 3 degrade G6P to pyruvate and CO_2, producing ATP, NADPH, and NADH. Mode 4 converts G6P into ribose-5-phosphate (R5P) and CO_2. It is of importance when the metabolic needs for reducing power for biosynthetic purposes and for R5P in nucleotide biosynthesis are balanced. In mode 5, five hexoses are converted into six pentoses. This is important when the need for R5P exceeds that for NADPH. Note that, depending on which type of utilization of R5P is considered, additional ATP consumption may occur in modes 4 and 5. The term "pentose phosphate cycle" applies to mode 6 best, because carbons are cycled several times before ending up in CO_2. Like modes 2 and 3, it is relevant when much more NADPH than R5P is required, but does not lead to NADH and ATP production. Mode 7 is a futile cycle formed by phosphofructokinase and fructose biphosphatase. The futile cycle is biochemically relevant because it generates additional heat in the organism—if required by environmental conditions. Further complexities include various modes that arise if genome-specific variations are analyzed or the reversibility of reactions is changed. Furthermore, from the start a smaller or larger network can be considered. Further details, descriptions, and conclusions on all these questions are found in Schuster et al. (2000).

An alternative to elementary mode analysis is to think of all possible metabolic fluxes as a type of space, the metabolic flux space. As in analytical geometry, one may then ask how this space is set up by vectors. These are the so-called basis vectors of flux space (Schilling and Palsson, 2000; Simpson et al., 1995; Fell, 1993). This has a problem, however: there are many possible solutions for the set of vectors defining the flux space. Nevertheless, very nice studies have been done using such flux balance studies—for example, investigating the comparative robustness of the *E. coli* metabolic network to supply cell growth over a wide range of different flux conditions (Edwards and Palsson, 2000a) or the effect of *E. coli* gene mutants (Edwards and Palsson, 2000b), as well as assessing the metabolic capabilities of *Haemophilus influenza* Rd (Schilling and Palsson, 2000). The different methods are further compared and discussed in detail in Schilling et al. (1999).

ANALYZING SPECIFIC PATHWAYS

We will now compare pathways in different genomes. This yields important information on their evolution, on pharmacological targets, and on biotechnological applications. We will discuss a pathway alignment approach in which three methods are combined: (1) analysis and comparison of biochemical data, (2) pathway analysis based on the concept of elementary modes, and (3) comparative genome analysis. For the last, the rapid increase of completely sequenced genomes is very helpful (e.g., see the available genomes under `http://www.tigr.org/tdb`). Comparing 17 completely sequenced genomes for their conservation of the glycolytic pathway could reveal a surprising plasticity of this very central pathway. This concerns both presence and absence of key enzymes as well as specific patterns of isoenzymes in different species. More deviations than previously thought from the textbook standard become apparent. Pathway alignment opens up new routes for potential pharmacological targets and reveals interesting bypasses such as the Entner-Doudoroff pathway. In a similar vein, archaean-, bacterial-, and parasite-specific adaptations can be rapidly identified and described.

Sequence alignment is a well-estalished tool for investigating and comparing nucleic acid and protein sequences from different species and for identifying characteristic gaps, insertions, and dissimilarities. The availability of full genomic sequences and the increasing amount of biochemical data open up higher-order possibilities for comparative analysis (Bork et al., 1998). Alignment of biochemical pathways from different species is an important step toward a more global physiological comparison.

Recognition of such differences is interesting for biotechnology (identification of alternative enzymes) and pharmacology (difference in drug targets). Biochemical data are extended and compared with elementary mode analysis of substrate fluxes, comparative genome analysis, and pathway alignment. Other valuable methods to consider and include for large-scale comparisons of genomic data and pathways include clusters of orthologous sequences (COGs; Tatusov et al., 2000) and the application of related enzyme clusters (Goto et al., 1997).

Pathway Alignment and Involved Bioinformatics Tools

The concept of this approach is to compare a range of organisms for one specific pathway, indicate the presence or absence of the involved enzymes by two different letters (or zeroes and ones), and then compare the enzyme sets in the same way as sequences are normally compared, marking well and less well conserved regions. Pathway alignment focuses on the biochemical capacity of the organisms compared. The utilization of complementary data and tools extends the biochemical approach and improves the predictive power of such efforts. A solid knowledge basis from literature and experimental data is established. Extensive use of sequence and genome analysis allows the characterization of the full repertoire of enzymes present in the analyzed organisms in regard to the analyzed pathway. The availability of full genomic sequences is exploited, and pathway fluxes and bypasses regarding the identified enzymes are tested for consistency by elementary mode analysis. Also, the determination of futile cycles in the different organism-specific enzyme sets is an indicator for allosteric regulation of key enzymes involved.

The collection of biochemical and genomic data required to establish presence or lack of enzymes prior to pathway alignment is greatly facilitated by the availability of databases on the World Wide Web (e.g., TIGR, `http://www.tigr.org/tdb/`; ExPASY, `http://www.expasy.ch/enzyme/index.html`; KEGG, `http://www.genome.ad.jp/kegg/kegg2.html`; WIT, `http://wit.mcs.anl.gov/WIT2/`). These systems are continously developed further (e.g., the orthologue group tables for KEGG; Bono et al., 1998). Differential metabolic display, another helpful pathway visualization and exploration tool, is based on petri nets to generate all pathways satisfying certain constraints (Kuffner et al., 2000; `http://cartan.gmd.de/ToPL/ign.html`). However, since databases are based on automated methods, including preformed pathway charts, these automated predictions are further refined for an accurate analysis.

Available completely sequenced genomes should be extensively cross-compared to better identify all encoded enzymes in these organisms, because in different genomes, different nonorthologous sequences may encode the same enzyme. Each of these families can be recognized only if at least one member has been biochemicaly characterized, a fact that stresses again the need for both genomic and biochemical databases and data. Genomic information is particularly valuable for organisms that are difficult to analyze biochemically. For genome cross-comparisons we recommend the extensive use of differential genome analysis, using assigned reading frames from one genome to compare its protein content against that of another genome and using application-specific chips for rapid sequence-to-sequence comparisons (Huynen et al., 1998). Gene duplications, replacement by unrelated sequences (nonorthologous displacement; see Bork et al., 1998), and gene neighborhoods also have to be taken into account. Phylogenetic analysis has to be applied in order to further analyze gene duplication events and clarify substrate specificities of the encoded enzymes.

Biochemical functions, sequence-based enzyme comparisons, and detailed enzyme specificities are made more accurate by a careful comparison of domain architecture and detection of duplication and gene displacement events, operon organization, and gene families. This is more exact than database overviews with a splendid amount of genome data that rely strongly on an automated assignment of enzyme functions by reciprocal gene similarities. As a medical application, identified enzymes from parasite-specific pathway variants may potentially be blocked pharmacologically.

Examples from Glycolysis

Applying pathway alignment for the analysis of glycolysis, the metabolic flow and corresponding enzymes from glucose 6-phosphate onward present a relatively well conserved region of the glycolytic pathway. However, in a number of organisms several glycolytic enzymes are missing, according to comparative sequence analysis and/ or biochemical data. For example, phosphofructokinase and aldolase activity are absent in *Mycoplasma hominis* (Pollack et al., 1997). Similarly, phosphofructokinase and pyruvate kinase seem to be lacking in *Helicobacter pylori*.

Different alternatives to processing glucose to pyruvate become apparent. In glycolysis two molecules of triose are derived from one hexose, and the energy yield is 2 moles of ATP per mole of glucose. The complete glycolytic pathway is present in *E. coli*. This is also the case in most eukaryotic (including human) cells. However, a first variation concerns the route of transfer of glucose into the cell. In contrast to higher organisms, the major *E. coli* pathway for glucose transport uses the PTS glucose transporter, which utilizes phosphoenolpyruvate as an energy-rich compound to directly phosphorylate glucose to glucose 6-phosphate.

The connections of pathways to other parts of metabolism, such as the connection of glycolysis to the pentose phosphate pathway, allow for further plasticity (e.g., the Entner-Doudoroff pathway is used instead of glycolysis in some bacteria; Danson and Hough, 1992). The ATP yield is only 1 mole per mole of glucose.

Furthermore, in *Methanococcus jannaschii*, all enzymes of the non-oxidative PPP were identified (Selkov et al., 1997). In contrast, the two dehydrogenases required for the oxidative PPP could be identified neither in the genome nor by biochemical assays. Glycolysis appears to be operative although hexokinase, phosphofructokinase, aldolase, and phosphoglycerate mutase could not be identified in the genome (Selkov et al., 1997). Assuming that these enzymes or other enzymes are present (for instance, there are some indications of ADP-dependent sugar kinases in archaea; Selkov et al., 1997; Kengen et al., 1995). The elementary modes for this system can then be calculated (Dandekar et al., 1999). One of these modes in fact allows the transformation of glucose 6-phosphate into ribose 5-phosphate, which is required for nucleotide biosynthesis. Thus, the oxidative PPP is not needed for converting hexoses into pentoses under steady-state conditions. This is in agreement with results of Pandolfi et al. (1995) on glucose 6-phosphate dehydrogenase deficiencies. A further tool for this type of analysis is enzyme hierarchies to find common patterns between pathways based on the EC classification (Tohsato et al., 2000)—which should be critically further classified by biochemical knowledge and comparisons.

Pathway alignment reveals species-specific pecularities and can become particularly powerful if coupled to metabolic engineering. Examples, would be engineering *Corynebacterium glutamicum* mutants to demonstrate two independent pathways not only for lysine but also for DL-diaminopimelate as an important building block of the murein cell wall of this bacterium (Sahm et al., 2000), or to better understand acetate production in *E. coli* by comparing it against other species and reducing its acetate synthesis by expressing a *B. subtilis* acetolactate synthase (Yang et al., 1999). Planning criteria for such projects that take results of metabolic flux analysis and a modular perturbation method into account are explained by Kholodenko et al. (1998); experimental suggestions for isotope labeling are given by Roscher et al. (2000).

PATHWAYS REVEALED BY GENOME ANALYSIS

Genome Annotation

Genome analysis is a complex and continuous process because the interpretation of the original genome sequence available after a major sequencing project and the experimental data available change during subsequent years. Therefore, before discussing additional techniques, we will illustrate pathway identification in the process of genome annotation and, with more information becoming available, subsequent reannotation.

For analysis the sequence has first to be available in electronic format. For a new sequencing project this is a rather cumbersome process and requires a great deal of technology. However, for studying and other purposes, many full genomic sequences and continuing public sequencing projects are deposited in public Web sites such as TIGR (`http://www.tigr.org/tdb/`).

In the next step, determining the position of potential reading frames, one has to keep in mind the need to distinguish between protein-coding and RNA-coding reading frames. This step is far from trivial. Important enzyme reading frames necessary to identify a complete pathway may, for example, escape detection because an extension of a reading frame escaped detection or an important enzyme activity is encoded on the opposite strand—not to mention notorious complications, particularly in higher eukaryotes, from alternative splicing or even editing of the mRNA.

If reading frames have been established and translated (note whether the proper genetic translation table has been used), a function has to be identified for the reading frame. Bioinformatics allows rapid comparison against existing data to achieve this. Thus protein reading frames can be assigned a putative function by sensitive sequence similarity comparisons. For detailed protocols and methodology the reader is referred to Bork et al. (1998) and Bork and Gibson (1996). However, to illustrate how this process allows completion of pathways, including identification of previously unnoted enzymatic activities, we will review some examples from a recent genome reannotation effort we conducted for *Mycoplasma pneumoniae* (Dandekar et al., 2000).

Pathway Completion by Reannotation

The reannotation of molecular functions can potentially provide some answers regarding higher levels of cellular interactions, such as metabolism (but also, for example, regarding pathogenicity factors).

Mycoplasma pneumoniae is a pathogen with a compact genome that causes atypical pneumonia in man. We analyzed the complete genome of *M. pneumoniae* five years after the original sequence was published (Himmelreich et al., 1996). The reannotation could rely on more data, in particular from molecular and biochemical analysis of *M. pneumoniae* during the intervening five years as well as better software for sequence analysis developed during that time. A well known example is PSI-BLAST, an iterative sequence search algorithm including sequence alignment information from similar sequences found during the search (Altschul et al., 1997).

In this way all individual reading frames of *M. pneunmoniae* were reexamined (Dandekar et al., 2000). We will review here some examples regarding better pathway identification or completion during the effort.

For example, the reading frame MPN547$_{(MP295)}$ [the new MP genome identifier is given first, followed by the (old MP genome identifier)] was originally annotated as a homologue of MG369, a conserved hypothetical protein of unknown function. This function was also indicated in the update of the *M. genitalium* genome sequence in December 1999. However, detailed sequence analysis in the meantime revealed more about the function. For example, there was a sequence similarity apparent in PSI-BLAST searches (see above). The expected odds of observing a similarity of this extent by chance was estimated to be 1 in 10 million (nevertheless, such expected values, E-values for short, are nontrivial to estimate and should be taken only as an indication for stronger confidence and for finding a biological confirmatory context for the much higher than chance similarity detected between the two sequences by the algorithm). There was similarity of the N-terminal 300 amino acids to experimentally characterized dihydroxyacetone kinases from different bacteria and fungi. Another way to show this relies on clusters of orthologous genes, COGs for short (Tatusov et al., 2000). The *M. pneumoniae* sequence can be shown to be a member of such a cluster of orthologous genes in which biochemically characterized members are dihydroxyacetone kinases.

In *M. pneumoniae*, the dihydroxyacetone kinase domain could yield ATP by transforming dihydroxyacetone phosphate and ADP into dihydroxyacetone and ATP. This finding should be interpreted in the context of other reading frames. Thus the predicted activity can be metabolically connected to the remaining phospholipid metabolism predicted for *Mycoplasma pneumoniae*. It could provide the necessary supply of dihydroxyacetone phosphate via MPN051$_{(MP103)}$ (glycerol 3-phosphate dehydrogenase reading frame, confirmed in reannotation). In addition, the remaining sequence of the MPN547$_{(MP295)}$ reading frame (total length 558 amino acids) may in addition regulate or add to this predicted enzyme activity.

To give another example, a ribulose uptake pathway is apparent from the reannotation. Small operons had been known previously for fructose (MPN078$_{(MP077)}$, to MPN079$_{(MP076)}$) and mannitol (MPN651$_{(MP191)}$ to MPN653$_{(MP189)}$). Ribulose was now found to be transported (MPN496$_{(MP346)}$, MPN494$_{(MP347)}$) and channeled via D-arabinose 6-hexulose 3-phosphate synthase (MPN493$_{(MP348)}$) and D-arabinose 6-hexulose 3-phosphate isomerase (MPN492$_{(MP349)}$) into fructose 6-phosphate and glycolysis. Of these proteins, MPN496 and MPN493 had not been functionally annotated before. MPN494 had been annotated as a hypothetical phosphotransferase. These new functional assignments can also be made apparent by integrating data from SWISS-PROT annotations (the well-known large database of protein reading frames) with further direct experimental data published for homologous proteins. In addition, we could now add the description of and direct experimental data (including mass-spectroscopy and sequence determination) for a small pentitol BC subunit of the ribulose transporter (MPN495$_{(MP346.1)}$) not included previously.

An example of a more complex and nonmetabolic pathway identified by genome reannotation in this study was the protein secretion system in *M. pneumoniae*. Himmelreich et al. (1996) noted that they had identified the trigger factor, DnaK, SRP and FtsY as well as SecA, whereas from the channel-forming proteins only SecY could be assigned, leaving the secretion pathway incomplete. Now using extensive sequence analysis methods, and integrating recent literature, bioinformtics, and biochemical data, the reannotation identified the reading frames similar to SecD, SecE, and SecG, yielding a new, more complete picture of this secretory pathway in *M. pneumoniae*. Since several pathogenicity factors are secreted, the respective protein channels are potential drug targets.

Similar analyses have been done on a number of different genomes, for example, a reannotation of the *Thermotoga maritima* genome (Kyrpides et al., 2000). Another aspect is the inclusion of structure and structural genomics, allowing analysis of structure-function relationships. This can be done either in a more summary way for whole genomes (e.g., comparing the spirochetes *T. pallidum* and *B. burgdorferi*. These lack, among other things, several parts of lipid biosynthesis and abundantly use, as is common in many other bacteria, TIM barrels and P-loop NTP hydrolases (Das et al., 2000). In addition, detailed examination of the enzymatic function, structure, and evolution of the different TIM barrel families, comparing all available data (Copley and Bork, 2000), shows widespread recruitment of enzymes between the central metabolic pathways.

PERSPECTIVE

We have given some examples of bioinformatic analysis of metabolic pathways, focusing on qualitative analyses; the detailed examples presented were mostly from our own work. The additional references given should make it clear that similar efforts have occurred in many laboratories around the world, and this chapter should serve as an incentive for further pathway analysis.

Pathway analysis has become more and more important because sequencing, and in particular the number of newly sequenced genomes, is rapidly increasing. Genome content is biologically meaningless unless interpreted in biological terms, and the metabolic pathways present an important part of this information. The description of the enzymatic capabilities based on functional genomics (e.g., via a tool such as pathway alignment) makes it possible to classify the great variety of organisms according to their different biochemical makeup. Medical aspects can now be tackled in unprecedented detail (e.g., analyzing metabolic changes in the liver, such as strong induction of the citric acid cycle and gluconeogenesis after major burns, which seem to be deranged mainly by a strong induction of the liver antioxidant defense pathways; Lee et al., 2000).

Missing enzyme activities and functional misassignment can happen all too easily (Kyrpides and Ouzounis, 1999). If possible, several tools (e.g., not only PSI-BLAST but also domain analysis, COGs, etc.) should be combined. Chapter 9 explains how several indications for protein function from genomics, in particular regarding interactions, are combined for better predictions, including involved pathways. Similarly, different lines of evidence should be combined (e.g., not only the genome sequence but also biological data). It should also be taken into account that any interpretation of biological data may change over time as more knowledge becomes available. Experimental data for metabolic pathway analysis are almost always incomplete. An interesting study by de Atauri et al. (2000) predicts the effect of uncertain boundary values in modeling a metabolic pathway. Such limitations and errors increase with integration to networks, cellular processes, or even the complete cell. In consequence, most models of this sort (see chapter 11) have to be interpreted with even more care.

The large-scale structure of metabolic networks has been investigated by several authors (Jeong et al., 2000; Fell and Wagner, 2000). A fascinating observation is that metabolic networks are connected as scale-free networks, meaning an exponential decrease in the number of more highly connected central metabolites. More complex organisms are more strongly connected than less complex ones. This improves adaptability to changing environmental conditions. The "diameter" of reactions (i.e., the number of enzymatic conversions connecting one enzymatic reaction of the network to a randomly chosen other one is kept close to the theoretically smallest possible number. In this way environmental perturbations rapidly spread over the network and are optimally buffered by the different enzyme activities. The effect is thus a "small world" way of connecting. Scale-free networks also apply to Internet hyperlinks or the U.S. power grid and to the "small world" everyday experience that human relations are surprisingly well connected via common friends or acquaintances.

The analysis of metabolic pathways is an overture to higher levels of simulation, not only the fascinating regulation of the metabolic pathways (level + 1, so to speak), but ultimately the whole cell (see chapter 11). A further valuable effort to bridge this gap while learning (or at least speculating) about prebiotic evolution is the analysis by Segre et al. (2000), modeling kinetically enhanced recruitment of simple amphiphilic molecules leading to evolving and splitting catalytic noncovalent assemblies.

The road to exploring the full metabolic network of the cell is open, and there are a number of very useful tools to analyze metabolic pathways. Nevertheless, there is still a very long way to travel before we can grasp the bewildering complexities of living cells.

REFERENCES

Altschul, S. F., Madden, T. L., Schaffer, A. A., Zhang, J., Zhang, Z., Miller, W., and Lipman, D. J. (1997). Gapped BLAST and PSI-BLAST: A new generation of protein database search programs. *Nucleic Acids Res.* 25: 3389–3402.

Bork, P., and Gibson, T. J. (1996). Applying motif and profile searches. *Methods Enzymol.* 266: 162–184.

Bono, H., Goto, S., Fujibuchi, W., Ogata, H., and Kanehisa, M. (1998). Systematic prediction of orthologous units of genes in the complete genomes. *Genome Inform. Ser. Workshop Genome Inform.* 9: 32–40.

Bork, P., Dandekar, T., Diaz-Lazcoz, Y., Eisenhaber, F., Huynen, M., and Yuan, Y. (1998). Predicting function: From genes to genomes and back. *J. Mol. Biol.* 283: 707–725.

Christensen, B., and Nielsen, J. (2000). Metabolic network analysis: A powerful tool in metabolic engineering. *Adv. Biochem. Eng. Biotechnol.* 66: 209–231.

Clarke, B. L. (1981). Complete set of steady states for the general stoichiometric dynamical system. *J. Chem. Phys.* 75: 4970–4979.

Copley, R. R., and Bork, P. (2000). Homology among (betaalpha) (8) barrels: Implications for the evolution of metabolic pathways. *J. Mol. Biol.* 303: 627–641.

Dandekar, T., et al. (2000). Re-annotating the *Mycoplasma pneumoniae* genome sequence: Adding value, function and reading frames. *Nucleic Acids Res.* 28: 3278–3288.

Dandekar, T., Schuster, S., Snel, B., Huynen, M., and Bork, P. (1999). Pathway alignment: Application to the comparative analysis of glycolytic enzymes. *Biochem. J.* 343: 115–124.

Danson, M. J., and Hough, D. W. (1992). The enzymology of archaebacterial pathways of central metabolism. *Biochem. Soc. Symp.* 58: 7–21.

Das, R., Hegyi, H., and Gerstein, M. (2000). Genome analyses of spirochetes: A study of the protein structures, functions and metabolic pathways in *Treponema pallidum* and *Borrelia burgdorferi*. *J. Mol. Microbiol. Biotechnol.* 2: 387–392.

de Atauri, P., Sorribas, A., and Cascante, M. (2000). Analysis and prediction of the effect of uncertain boundary values in modeling a metabolic pathway. *Biotechnol. Bioeng.* 68: 18–30.

Edwards, J. S., and Palsson, B. O. (2000a). Robustness analysis of the *Escherichia coli* metabolic network. *Biotechnol Prog.* 16: 927–939.

Edwards, J. S., and Palsson, B. O. (2000b). Metabolic flux balance analysis and the *in silico* analysis of *Escherichia coli* K-12 gene deletions. *BMC Bioinformatics* 1: 1.

Fell, D. A. (1993). In S. Schuster, M. Rigoulet, R. Ouhabi, and J.-P. Mazat (eds.), *Modern Trends in Biothermokinetics*. New York: Plenum Press, pp. 97–101.

Fell, D. A., and Wagner, A. (2000). The small world of metabolism. *Nat. Biotechnol.* 18: 1121–1122.

Galperin, M. Y., and Brenner, S. E. (1998). Using metabolic pathway databases for functional annotation. *Trends Genetics* 14: 332–333.

Goto, S., Bono, H., Ogata, H., Fujibuchi, W., Nishioka, T., Sato, K., and Kanehisa, M. (1997). Organizing and computing metabolic pathway data in terms of binary relations. *Pac. Symp. Biocomput.* 1997: 175–186.

Hatzimanikatis, V. (1999). Nonlinear metabolic control analysis. *Metab. Eng.* 1: 75–87.

Himmelreich, R., Hilbert, H., Plagens, H., Pirkl, E., Li, B. C., and Herrmann, R. (1996). Complete sequence analysis of the genome of the bacterium *Mycoplasma pneumoniae. Nucleic Acids Res.* 24: 4420–4449.

Huynen, M., Dandekar, T., and Bork, P. (1998). Differential genome analysis: Rapid identification of genes encoding species-specific phenotypes. *FEBS Lett.* 426: 1–5.

Jeong, H., Tombor, B., Albert, R., Oltvai, Z. N., and Barabasi, A. L. (2000). The large-scale organization of metabolic networks. *Nature* 407: 651–654.

Kengen, S. W., Tuininga, J. E., de Bok, F. A., Stams, A. J., and de Vos, W. M. (1995). Purification and characterization of a novel ADP-dependent glucokinase from the hyperthermophilic archaeon *Pyrococcus furiosus. J. Biol. Chem.* 270: 30453–30457.

Kholodenko, B. N., Cascante, M., Hoek, J. B., Westerhoff, H. V., and Schwaber, J. (1998). Metabolic design: How to engineer a living cell to desired metabolite concentrations and fluxes. *Biotechnol. Bioeng.* 59: 239–247.

Kuffner, R., Zimmer, R., and Lengauer, T. (2000). Pathway analysis in metabolic databases via differential metabolic display. *Bioinformatics* 16: 825–836.

Kyrpides, N. C., and Ouzounis, C. A. (1999). Whole-genome sequence annotation: "Going wrong with confidence." *Mol. Microbiol.* 32: 886–887.

Kyrpides, N. C., Ouzounis, C. A., Iliopoulos, I., Vonstein, V., and Overbeek, R. (2000). Analysis of the *Thermotoga maritima* genome combining a variety of sequence similarity and genome context tools. *Nucleic Acids Res.* 28: 4573–4576.

Laufs, S., Kim, S. H., Kim, S., Blau, N., and Thony, B. (2000). Reconstitution of a metabolic pathway with triple-cistronic IRES-containing retroviral vectors for correction of tetrahydrobiopterin deficiency. *J. Gene Med.* 2: 22–31.

Lee, K., Berthiaume, F., Stephanopoulos, G. N., Yarmush, D. M., and Yarmush, M. L. (2000). Metabolic flux analysis of postburn hepatic hypermetabolism. *Metab. Eng.* 2: 312–327.

Pandolfi, P. P., Sonati, F., Rivi, R., Mason, P., Grosveld, F., and Luzzatto, L. (1995). Targeted disruption of the housekeeping gene encoding glucose 6-phosphate dehydrogenase (G6PD): G6PD is dispensable for pentose synthesis but essential for defense against oxidative stress. *EMBO J.* 14: 5209–5215.

Pfeiffer, T., Sánchez-Valdenebro, I., Nuño, J. C., Montero, F., and Schuster, S. (1999). METATOOL: For studying metabolic networks. *Bioinformatics* 15(3): 251–257.

Pollack, J. D., Williams, M. V., and McElhaney, R. N. (1997). The comparative metabolism of the mollicutes (Mycoplasmas): the utility for taxonomic classification and the relationship of putative gene annotation and phylogeny to enzymatic function in the smallest free-living cells. *Crit. Rev. Microbiol.* 23: 269–354.

Roscher, A., Kruger, N. J., and Ratcliffe, R. G. (2000). Strategies for metabolic flux analysis in plants using isotope labelling. *J. Biotechnol.* 77: 81–102.

Sahm, H., Eggeling, L., and de Graaf, A. A. (2000). Pathway analysis and metabolic engineering in *Corynebacterium glutamicum*. *Biol. Chem.* 381: 899–910.

Schilling, C. H., Letscher, D., and Palsson, B. O. (2000). Theory for the systemic definition of metabolic pathways and their use in interpreting metabolic function from a pathway-oriented perspective. *J. Theor. Biol.* 203: 229–248.

Schilling, C. H., and Palsson, B. O. (2000). Assessment of the metabolic capabilities of *Haemophilus influenzae* Rd through a genome-scale pathway analysis. *J. Theor. Biol.* 203: 249–283.

Schilling, C. H., Schuster, S., Palsson, B. O., and Heinrich, R. (1999). Metabolic pathway analysis: Basic concepts and scientific applications in the post-genomic era. *Biotechnol. Prog.* 15: 296–303.

Schuster, S., and Hilgetag, C. (1994). On elementary flux modes in biochemical reaction systems at steady state. *J. Biol. Syst.* 2: 165–182.

Schuster, R. and Schuster, S. (1993) Refined algorithm and computer program for calculating all non-negative fluxes admissible in steady states of biochemical reaction systems with or without some flux rates fixed. *Comp. Appl. Biosci.* 9: 79–85.

Schuster, S., Fell, D., and Dandekar, T. (2000). A general definition of metabolic pathways useful for systematic organization and analysis of complex metabolic networks. *Nat. Biotechnol.* 18: 326–332.

Segre, D., Ben-Eli, D., and Lancet, D. (2000). Compositional genomes: Prebiotic information transfer in mutually catalytic noncovalent assemblies. *Proc. Natl. Acad. Sci. USA* 97: 4112–4117.

Selkov, E., Maltsev, N., Olsen, G. J., Overbeek, R., and Whitman, W. B. (1997). A reconstruction of the metabolism of *Methanococcus jannaschii* from sequence data. *Gene* 197: GC11–26.

Seressiotis, A., and Bailey, J. E. (1986). MPS: An algorithm and data base for metabolic pathway synthesis. *Biotechnol. Lett.* 8: 837–842.

Simpson, T. W., Colón, G. E., and Stephanopoulos, G. (1995). Two paradigms of metabolic engineering applied to amino acid biosynthesis. *Biochem. Soc. Trans.* 23: 381–387.

Stephanopoulos, G. (1999). Metabolic fluxes and metabolic engineering. *Metab. Eng.* 1: 1–11.

Tatusov, R. L., Galperin, M. Y., Natale, D. A., and Koonin, E. V. (2000). The COG database: A tool for genome-scale analysis of protein functions and evolution. *Nucleic Acids Res.* 28: 33–36.

Tohsato, Y., Matsuda, H., and Hashimoto, H. (2000). A multiple alignment algorithm for metabolic pathway analysis using enzyme hierarchy. *ISMB* 8: 376–383.

Vallino, J. J., and Stephanopoulos, G. (1993). Metabolic flux distributions in *Corynebacterium glutamicum* during growth and lysine overproduction. *Biotechnol. Bioeng.* 41: 633–646.

Womble, D. D., and Rownd, R. H. (1986). Regulation of lambda dv plasmid DNA replication. A quantitative model for control of plasmid lambda dv replication in the bacterial cell division cycle. *J. Mol. Biol.* 191: 367–382.

Yang, Y. T., Aristidou, A. A., San, K. Y., and Bennett, G. N. (1999). Metabolic flux analysis of *Escherichia coli* deficient in the acetate production pathway and expressing the *Bacillus subtilis* acetolactate synthase. *Metab. Eng.* 1: 26–34.

SUGGESTED READING

Cornish-Bowden, A., and Cardenas, M. L. (2000). From genome to cellular phenotype—a role for metabolic flux analysis? *Nat. Biotechnol.* 18: 267–268.

Dandekar, T., Schuster, S., Snel, B., Huynen, M., and Bork, P. (1999). Pathway alignment: Application to the comparative analysis of glycolytic enzymes. *Biochem. J.* 343: 115–124.

Fell, D. A. (1997). *Understanding the Control of Metabolism.* London: Portland Press.

Schuster, S. Dandekar, T., and Fell, D. A. (1999). Detection of elementary flux modes in biochemical networks: A promising tool for pathway analysis and metabolic engineering. *Trends Biotechnol.* 17: 53–60.

URLs FOR RELEVANT SITES

BRENDA, A collection of enzyme functional data including kinetic parameters and specificity. `http://www.brenda.uni-koeln.de/`

A detailed description of an algorithm for computing elementary flux modes. `http://bms-mudshark.brookes.ac.uk/algorithm.pdf`

Download site for EMPATH and METATOOL, two implementations of the elementary flux mode calculation algorithm described in this chapter. `ftp://bmshuxley.brookes.ac.uk/pub/mca/software/ibmpc`

ENZYME, A database containing information relative to the nomenclature and classification of enzymes. `http://www.expasy.ch/enzyme/index.html`

KEGG (Kyoto Encyclopedia of Genes and Genomes), a collection of computerized metabolic pathways with extensive links to other databases. `http://www.genome.ad.jp/kegg`

The Institute for Genomic Research (TIGR) provides a list of genome sequencing projects—a wealth of genomic information. `http://www.tigr.org/tdb/`

Further explanation of the program METATOOL. `http://www.biologie.hu-berlin.de/biophysics/Theory/tpfeiffer/metatool.html`

WIT, a project seeking to reconstruct metabolism for sequenced genomes that contains many metabolic pathway maps and enzyme lists. `http://wit.mcs.anl.gov/WIT2/`

11 Toward Computer Simulation of the Whole Cell

Masaru Tomita

It is still an open question whether or not computer simulation of a whole cell by modeling all its metabolic pathways and other machineries is feasible. Yet the time has come that we can begin working toward this ultimate goal. The E-CELL project (Tomita et al., 1999) was launched in 1996 at Keio University in order to model and simulate various cellular processes, with the ultimate goal of simulating the cell as a whole. The E-CELL simulation system is a generic software package for cell modeling, and its first version was completed in 1997. E-CELL simulates the behavior of a model cell by computing interactions among molecular species it contains. E-CELL allows the user to define chemically discrete compartments such as membranes, chromosomes, and cytoplasm. The total amounts of substances in all of the cellular compartments comprise the internal representation of the cell state, which can be monitored and/or manipulated by means of various graphical user interfaces (figure 11.1).

The E-CELL system enables us to model not only metabolic pathways but also other higher-order cellular processes, such as protein synthesis and membrane transport, within the same framework. E-CELL also attempts to provide a framework for higher-order cellular phenomena such as the gene regulation network, DNA replication, and other events in the cell cycle. These various processes can then be integrated into a single simulation model. A major challenge of the project is to develop a method of modeling that is sufficiently robust to accommodate a realistically large scale model consisting of hundreds and thousands of processes with drastic differences in behavior and time scale among them.

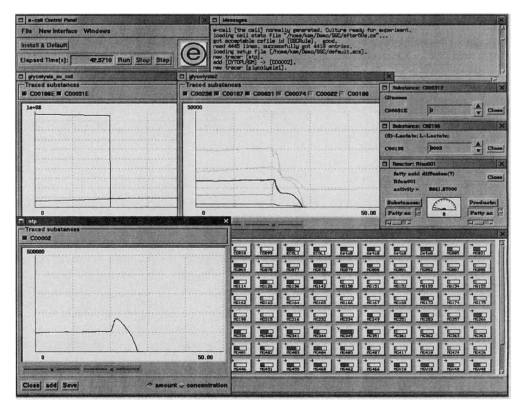

Figure 11.1 Screen dump of the E-CELL simulation system. Various graphical user interfaces make it possible not only to monitor metabolism of the virtual cell, but also to conduct virtual experiments in silico.

Using the E-CELL system, we have successfully constructed a virtual cell with 127 genes sufficient for "self-support" (Tomita et al., 1999). The gene set was selected from the genome of *Mycoplasma genitalium*, the organism having the smallest known genome. The set includes genes for transcription, translation, the glycolysis pathway for energy production, membrane transport, and the phospholipid biosynthesis pathway for membrane structure.

The rest of this section will review previous work related to computer simulation of cellular processes. Following sections describe the E-CELL simulation system and present the first "virtual cell" with 127 genes.

The following cell models are currently being constructed using the E-CELL system:

- Kinetic model of a human erythrocyte
- Signal transduction for bacterial chemotaxis
- Gene regulation network for E. *coli*'s *lac* and *ara* operons and the lytic-lysogenic switch network of bacteriophage lambda
- Energy metabolism in mitochondria.

These models are presented in later sections of the chapter.

Many attempts have been made to simulate molecular processes in both cellular and viral systems. Perhaps the most active area of cellular simulation is the kinetics of metabolic pathways. Software packages for quantitative simulation of cellular processes, based on numerical integration of rate equations, have been developed. KINSIM (Barshop et al., 1983; Dang and Frieden, 1997), MetaModel (Cornish-Bowden and Hofmeyr, 1991), SCAMP (Sauro, 1993), and MIST (Ehlde and Zacchi, 1995) deal with steady states and their metabolic control analysis (MCA) coefficients. GEPASI (Mendes, 1993, 1997) calculates steady states as well as reaction time behavior and characterizes the steady state with MCA. SIMFIT (Holzhutter and Colosimo, 1990) and KINSIM (Barshop et al., 1983; Dang and Frieden, 1997) integrate the simulation of dynamic models and the parameter fitting program. V-Cell is a solver for nonlinear PDE/ODE/algebraic systems, and can represent the cellular geometry. Dbsolve (Goryanin et al., 1999) combines continuation and bifurcation analysis. Several groups have proposed and analyzed gene regulation and expression models by simulation (Meyers and Friedland, 1984; Koile and Overton, 1989; Karp, 1993; Arita et al., 1994; McAdams and Shapiro, 1995). The cell division cycle (Tyson, 1991; Novak and Tyson, 1995) and signal transduction mechanisms (Bray et al., 1993) have also been areas of research for biological modeling and simulation.

Previous work in biochemical and genetic simulations has usually limited its models to focus on only one of the several levels of the time-scale hierarchy in cellular processes. Bridging the gaps between the various levels of this hierarchy is an extremely challenging problem that has yet to be adequately addressed. We present a step toward integrative simulation of large-scale cellular processes.

THE E-CELL SOFTWARE ARCHITECTURE

The E-CELL software employs a *structured substance-reactor model* to construct and simulate the cell model. The structured substance-reactor model consists of three classes of objects—*Substance, Reactor,* and *System*—representing molecular species, reactions, and functional/physical compartments, respectively. To model chromosomes and other genetic materials, the system also has sophisticated data structures such as *Genome, GenomicElement,* and *Gene.*

Assuming rapid equilibrium in each compartment, the state of the cell at a point in time and its dynamics are represented as a set of concentration vectors and reaction kinetics. Therefore, simulation of the cell can be viewed as integration of ordinary differential equations (ODEs). The numerical integration is performed in distributed form with *Stepper* and *Integrator* class objects in the system. With this design, the system can be extended to handle multicompartment, multischeme, and multiphase numerical integration. In other words, time-step sizes and integration algorithms can vary among different compartments, and therefore parallel computation can be applied to cope with the *stiffness* problem. The problem of stiffness occurs in a system like a cell where there are very different scales of the time variable on which the dependent variables are changing. The stiffness often results in various kinds of numerical errors, some of which are fatal for the simulation.

The E-CELL system is written in C++ and designed with the object-oriented MVC (model, view, and control) model, which facilitates independent development of the reaction rules (Model), user interfaces (View), and simulation engine (Control). The kinetics of each reaction is encapsulated into the Reactor class objects. In addition to general reaction schemes provided by the system, task-specific reaction schemes can be defined by a user without detailed knowledge of implementation of the other parts of the system.

The E-CELL simulation environment consists of several software components: the core simulation system (E-CELL system), the E-CELL manager, the rule file compiler, and other data converters/processors. Besides the system-specific E-CELL rule format, the user can write simulation rules using any commercially/publicly available spreadsheet software. The rules are compiled into .eri (E-CELL Rule Intermediate) format, which can be interpreted by the simulation system. The core system reads a compiled rule, starts simulation, and outputs

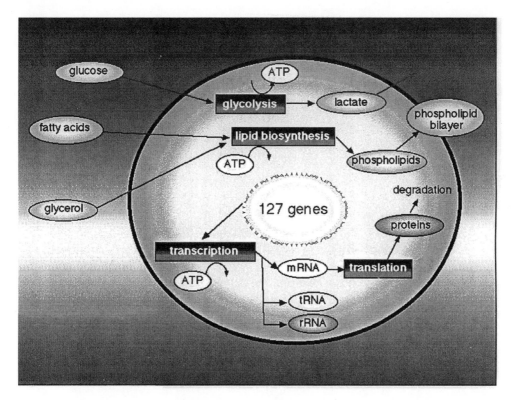

Figure 11.2 The self-surviving cell model with 127 genes. It uptakes glucose and excretes lactate, generating ATP to be used to synthesize proteins. Since proteins degrade over time, the cell has to keep synthesizing proteins in order to sustain its life.

logged data. The simulation process can be automated using E-CELL Script language (.ecs). The process of rule writing, simulation, and analysis is managed by the E-CELL manager.

THE SELF-SUPPORTING CELL MODEL

Using the E-CELL system, we have successfully constructed a virtual cell with 127 genes sufficient for "self-support" (figure 11.2). The gene set was selected from the genome of *Mycoplasma genitalium*, the organism having the smallest known genome. The set includes genes for transcription, translation, the glycolysis pathway for energy production, membrane transport, and the phospholipid biosynthesis pathway for membrane structure.

The hypothetical cell we have modeled uptakes glucose from the culture medium, using a phosphotransferase system; generates ATP by catabolizing glucose to lactate by glycolysis and fermentation pathways; and exports lactate out of the cell. Since enzymes and other proteins are modeled to degrade spontaneously over time, they must be constantly synthesized in order for the cell to sustain life. The protein synthesis is implemented by modeling the molecules necessary for transcription and translation: RNA polymerase, ribosomal subunits, rRNAs, tRNAs, and tRNA ligases. The cell also uptakes glycerol and fatty acid, and produces phosphatidyl glycerol for membrane structure, using a phospholipid biosynthesis pathway.

The E-CELL interfaces provide a means of conducting experiments *in silico*. For example, we can starve the cell by draining glucose from the culture medium. The cell will eventually die because it runs out of ATP. If glucose is added back, the cell may or may not recover, depending on the length of starvation. We can also kill the cell by knocking out an essential gene—for example, that for protein synthesis. The cell will become unable to synthesize proteins, and all enzymes will eventually disappear due to spontaneous degradation.

THE HUMAN ERYTHROCYTE MODEL

The human erythrocyte has been well studied since the 1970s, and extensive biochemical data on its enzymes and metabolites have been accumulated (Joshi and Palsson, 1989–1990; Ni and Savageau, 1996a, 1996b; Lee and Palsson, 1990, 1992; Tanaka and Paglia, 1995). The cell uptakes glucose from the environment and processes it through the glycolysis pathway, generating ATP molecules for other cellular metabolism. The ATP molecules are consumed mostly for cation transport in order to keep the electroneutrality and osmotic balance. The cell also has several other pathways, such as nucleotide metabolism and the pentose phosphate pathway.

The first prototype of our cell model (basic model), which was completed in March 1999, consists of the glycolysis pathway, the pentose phosphate pathway, and nucleotide metabolism (figure 11.3). Parameters of their kinetic equations are based on experimental data found in the literature. We obtained the steady state with this model and now set out to analyze the consequences of enzyme deficiencies.

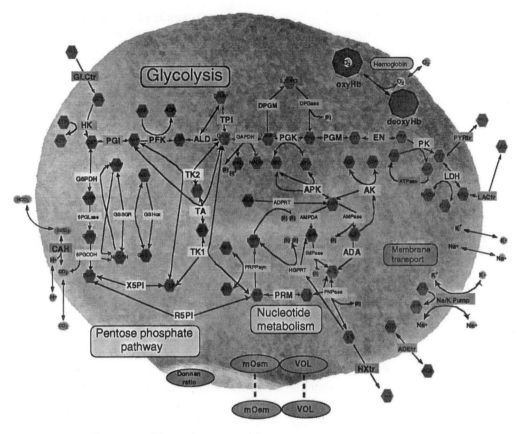

Figure 11.3 The erythrocyte model.

The model has been extended in various ways since then, and the latest version is capable of simulating the pH dependence of enzymes, osmotic balance, electroneutrality, and oxygen and carbon dioxide transportation by hemoglobin. The current model can also represent the hemolysis and the destructive processing of abnormal erythrocytes in the spleen. Since hemogrobin and carbon dioxide have a buffering effect for pH change, and osmotic balance can absorb quantitative changes of metabolic intermediates, these functions will probably enhance the robustness and tolerance of the model. We are currently analyzing the effect of this expansion on robustness and tolerance.

Since the discovery of pyruvate kinase deficiency (Tanaka and Paglia, 1995), erythroenzymopathies associated with hereditary hemo-

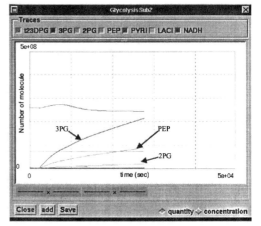

Figure 11.4 Simulation of pyruvate kinase deficiency. Pi, inorganic phosphates; Mg, free magnesium; tAMP, total AMP; tADP, total ADP; tATP, total ATP; T23 DPG, total 2,3-diphosphoglycerate; 3PG, 3-phosphoglycerate; 2G, 2-phosphoglycerate; PEP, phosphoenol pyruvate; LACi, lactate.

lytic anemia have been extensively investigated. We are currently trying to reconstruct the erythrocyte model with these deficiencies. Simulation experiments using the basic model of the human erythrocyte were carried out. Pyruvate kinase activity was set to zero at 2000 seconds. The ATP production rate became lower, and ATP was eventually exhausted. Increases of 2-phosphoglycerate, 3-phosphoglycerate, and phosphoenolpyruvate were observed (figure 11.4). These phenomena were reported in human erythrocytes from patients with PK (pyruvate kinase) deficiency. We plan to use the kinetic parameters of the PK mutant to obtain the steady state of the defective erythrocyte.

SIGNAL TRANSDUCTION FOR BACTERIAL CHEMOTAXIS

Chemotaxis is the orientation of an organism in relation to the presence of a particular chemical. The chemotactic response of *E. coli* depends on the ability to modulate the flagellar motor in response to external stimuli. The flagellar motor has a switch protein that interacts with proteins called CheY. The CheY protein in a phosphorylated form is known to bind with the motor up to 20 times more frequently than that in an

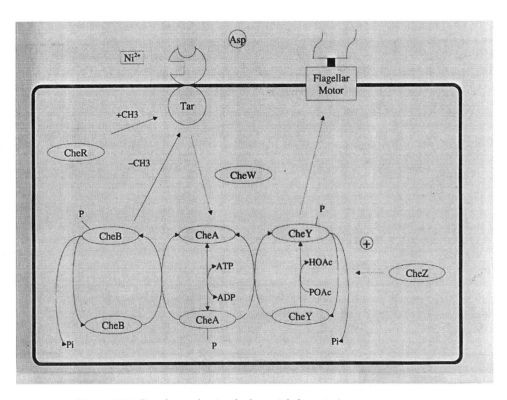

Figure 11.5 Signal transduction for bacterial chemotaxis.

unphosphorylated form. In this way, the phosphorylation controls bacterial response to external stimuli by regulating the binding affinity of the CheY protein and the motor switch.

The external stimulus bound to a bacterial receptor works as a initial control element of the cytoplasmic phosphorylation cascade. If the stimulus is an attractant, the phosphorylation flow is suppressed and the motor rotates counterclockwise, resulting in smooth swim.

We are currently modeling the phosphorylation cascade for bacterial (*E. coli*) chemotaxis in which the attractant is aspartate (figure 11.5). The E-CELL system allows us to perform virtual experiments on the bacterial chemotaxis model. For example, mutants of the model can be created by making a particular protein always zero. The simulated behavior of each mutant, as well as the wild type, can then be compared with the results of laboratory experiments, using indexes such as swimming direction, response time, and phosphate flux. We also plan to compare our results with other simulation systems (Bray et al., 1993; Bray and Bourret, 1995).

GENE REGULATION NETWORK

This section presents a general framework for modeling gene expression using the E-CELL system. Using the framework, we modeled and simulated the following three gene regulation systems: the *lac* operon of *E. coli*, the *ara* operon of *E. coli*, and the lytic-lysogenic switch network in bacteriophage lambda.

We had previously constructed a detailed model (Hashimoto et al., 1998) of the gene expression system as a part of the "self-supporting cell model." The previous model faithfully reflected the gene expression system of *M. genitalium*, involving more than 100 objects including subunits, factors, amino acids, nucleotides, and tRNAs, and their ligases. However, in order to simulate complex gene regulation systems with a large number of genes, a more abstract and simpler model is desired for the sake of efficiency. The gene expression system used in this work consists of four basic elements: (1) production of RNAs and proteins, (2) regulation of expression by various factors, (3) time delay between regulation and production, and (4) increase of the sigmoidal curve of product over time. All gene expressions are reduced to these four elements.

E. coli's *lac* Operon and *ara* Operon

Based on the framework described above, we have modeled gene regulation systems of the *lac* operon and the *ara* operon in *E. coli*. We are currently integrating the *lac* operon model with the glycolysis pathway (figure 11.6), so that we can simulate *E. coli*'s sugar metabolism and regulation with lactose and glucose. The *ara* operon model is being integrated with the pentose phosphate cycle. Results of these simulations will be compared and analyzed with experimental data reported in the literature (Kepes, 1969).

Lytic-Lysogenic Switch Network of Bacteriophage Lambda

As a more complex gene regulation system, we are modeling the lytic-lysogenic switch network of bacteriophage lambda. The model of the

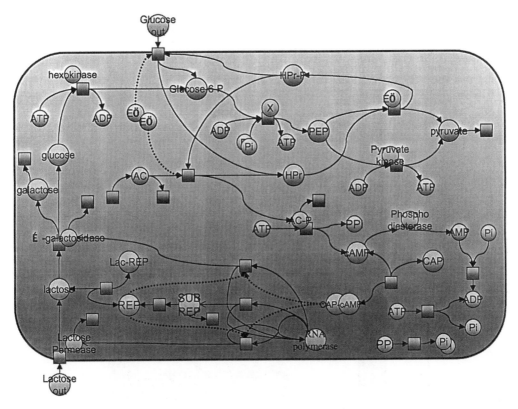

Figure 11.6 Integrative simulation of energy metabolism and gene expression.

lytic-lysogenic switch network includes the effects of glucose depriva-
tion, ultraviolet light, and multiplicity of infection.

Since the E-CELL system allows easy substitution of individual reactions, several variations of gene expression models of bacteriophage
lambda reported in the literature (McAdams and Shapiro, 1995;
Heidtke and Schulze-Kremer, 1998) will be constructed and compared
using the E-CELL system.

MITOCHONDRIA MODEL

We have also been developing kinetic models of various mitochondrial
metabolisms, including gene expression, energy metabolism, fatty acid
metabolism (beta oxidation), inner-membrane metabolite carriers, and
protein carriers (figure 11.7). In this section, we report the model of

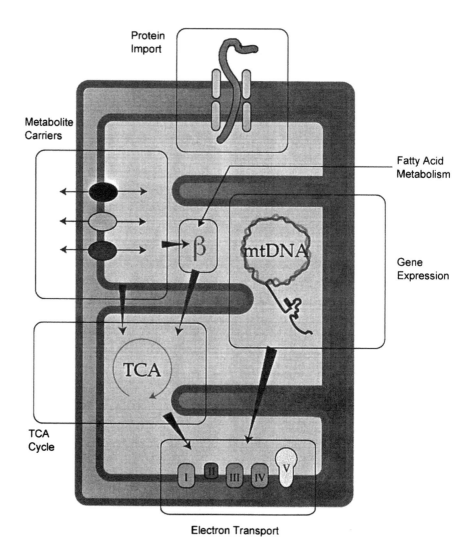

Figure 11.7 The mitochondria model consists of gene expression system, protein transport, metabolite carriers, TCA cycle, electron transport, and fatty acid metabolism. Its genome has 37 genes, and the total of 30 enzymatic reactions are modeled.

Masaru Tomita

pathways for energy metabolism that was recently completed using the E-CELL system.

A mitochondrion has three major energy metabolic pathways: electron transport, TCA cycle, and beta oxidation. The basic steps of energy metabolism are listed below.

· Inner-membrane metabolite carriers transport pyruvate, fatty acid, ADP, and Pi.

· Acetyl-CoA is produced from pyruvate by pyruvate dehydrogenase.

· Acetyl-CoA is made from fatty acid by beta oxidation.

· The TCA cycle oxidizes acetyl-CoA to produce NADH, the substrate of electron transport.

· The four enzymes of electron transport oxidize NADH.

· At the end of electron transport, ATP synthase generates ATP.

There are five enzymatic reactions for electron transport, eleven enzymatic reactions, and nine other reactions for the TCA cycle, five enzymatic reactions for beta oxidation, and eight enzymatic reactions for the metabolite carrier system. All the enzymatic reactions are modeled on the basis of kinetic parameters found in the literature.

We will continue to enhance this mitochondrial model, adding other major part of metabolism, including gene expression, fatty acid metabolism, and DNA replication. Our eventual goal is to apply this model to pathological analyses of mitochondrial diseases such as Leigh syndrome, mitochondrial myopathy and encephalopathy, lactic acidosis, and strokelike episodes.

CONCLUDING REMARKS

The E-CELL system has been available since July 1999 from our Web site (http://www.e-cell.org). Some of the models described in this chapter can be downloaded from this Web site. Our simulation work with E-CELL has shown that large-scale modeling of various cellular metabolisms using the same framework appears feasible.

One of the major problems in constructing large-scale cell models is lack of quantitative data. Most of the biological knowledge available is qualitative (e.g., functions of genes, pathway maps, which proteins interact with what, etc.). For simulation, however, quantitative data such as concentrations of metabolites and enzymes, flux rates, kinetic parameters, and dissociation constants are essential. A major challenge is to develop high-throughput technologies for measurement of inner-cellular metabolites. A great deal of data for a variety of cell states can then be collected with the technologies to construct quantitative models, and the models can be refined iteratively until the simulation results match the data.

For this new type of simulation-oriented biology, Keio University has established the Institute for Advanced Biosciences (`http://www.bioinfo.sfc.keio.ac.jp/IAB/`). The institute consists of centers for metabolome research, bioinformatics, and genome engineering. The ultimate goal of this international research institute is to construct a whole-cell model *in silico*, based on a large amount of data generated by high-throughput metabolome analyses, then design a novel genome based on the computer simulation and create real cells with the novel genome by means of genome engineering.

As mentioned at the beginning of this chapter, whether or not complete living cells can be modeled and simulated at the molecular level is still an open question. However, the rapidly accumulating information on cellular metabolism increases the likelihood that whole-cell simulation of real organisms will be feasible in the near future.

ACKNOWLEDGMENTS

I would like to thank all the E-CELL project members of the Laboratory for Bioinformatics, Keio University, including Kenta Hashimoto, Koichi Takahashi, Yoichi Nakayama, Yuri Matsuzaki, Fumihiko Miyoshi, Katsuyuki Yugi, and Yusuke Saito. The E-CELL project is funded in part by Japan Science and Technology Corporation and a grant-in-aid for scientific research on priority areas from the Ministry of Education, Science, Sports, and Culture of Japan.

REFERENCES

Arita, M., Hagiya, M., and Shiratori, T. (1994). GEISHA SYSTEM: An environment for simulating protein interaction. In T. Takagi (ed.), *Proceedings, Genome Informatics Workshop 1994*. Tokyo: Universal Academy Press, pp. 81–89.

Barshop, B. A., Wrenn, R. F., and Frieden, C. (1983). Analysis of numerical methods for computer simulation of kinetic processes: Development of KINSIM—a flexible, portable system. *Anal. Biochem.* 130: 134–145.

Bray, D., and Bourret, R. B. (1995). Computer analysis of the binding reactions leading to a transmembrane receptor-linked multiprotein complex involved in bacterial chemotaxis. *Mol. Biol. Cell* 6: 1367–1380.

Bray, D., Bourret, R. B., and Simon, M. I. (1993). Computer simulation of the phosphorylation cascade controlling bacterial chemotaxis. *Mol. Biol. Cell* 4: 469–482.

Cornish-Bowden, A., and Hofmeyr, J. H. (1991). MetaModel: A program for modeling and control analysis of metabolic pathways on the IBM PC and compatibles. *Comput. Appl. Biosci.* 7: 89–93.

Dang, Q., and Frieden, C. (1997). New PC versions of the kinetic-simulation and fitting programs, KINSIM and FITSIM. *Trends Biochem. Sci.* 22 (8): 317.

Ehlde, M., and Zacchi, G. (1995). MIST: A user-friendly metabolic simulator. *Comput. Appl. Biosci.* 11: 201–207.

Goryanin, I., Hodgman, T. C., and Selkov, E. (1999). Mathematical simulation and analysis of cellular metabolism and regulation. *Bioinformatics* 15: 749–758.

Hashimoto, K., Miyoshi, F., Shimizu, T. S., Satoyoshi, T., and Tomita, M. (1998). Modelling of transcription and DNA replication using the E-CELL simulation system. In *Genome Informatics 1998*. Tokyo: Universal Academy Press, pp. 244–245.

Heidtke, K. R., and Schulze-Kremer, S. (1998). Design and implementation of a qualitative simulation model of lambda phage infection. *Bioinformatics* 14 (1): 81–91.

Holzhutter, H. G., and Colosimo, A. (1990). SIMFIT: A microcomputer software toolkit for modelistic studies in biochemistry. *Comput. Appl. Biosci.* 6: 23–28.

Joshi, A., and Palsson, B. O. (1989–1990). Metabolic dynamics in the human red cell. Part I—A comprehensive kinetic model. *J. Theor. Biol.* 141, no. 4 (Dec. 19, 1989): 515–528; Part II—Interactions with the environment. *J. Theor. Biol.* 141, no. 4 (Dec. 19, 1989): 529–545; Part III—Metabolic reaction rates. *J. Theor. Biol.* 142, no. 1 (Jan. 9, 1990): 41–68; Part IV—Data prediction and some model computations. *J. Theor. Biol.* 142, no. 1 (Jan. 9, 1990): 69–85.

Karp, P. D. (1993). A qualitative biochemistry and its application to the regulation of the tryptophan operon. In L. Hunter (ed.), *Artificial Intelligence and Molecular Biology*. Menlo Park Calif.: AAAI Press; Cambridge, Mass.: MIT Press, pp. 289–324.

Kepes, A. (1969). Transcription and translation in the lactose operon of *Escherichia coli* studied by in vivo kinetics. *Prog. Biophys. Mol. Biol.* 19 (1): 199–236.

Koile, K., and Overton, G. C. (1989). A qualitative model for gene expression. In *Proceedings of the 1989 Summer Computer Simulation Conference*. pp. 415–421.

Lee, I. D., and Palsson, B. O. (1990). A comprehensive model of human erythrocyte metabolism: Extensions to include pH effects. *Biomed. Biochem.* 49 (8–9): 771–789.

Lee, I. D., and Palsson, B. O. (1992). A Macintosh software package for simulation of human red blood cell metabolism. *Comput. Methods Programs Biomed.* 38 (4): 195–226.

McAdams, H. H., and Shapiro, L. (1995). Circuit simulation of genetic networks. *Science* 269: 650–656.

Mendes, P. (1993). GEPASI: A software package for modeling the dynamics, steady states and control of biochemical and other systems. *Comput. Appl. Biosci.* 9: 563–571.

Mendes, P. (1997). Biochemistry by numbers: Simulation of biochemical pathways with Gepasi 3. *Trends Biochem. Sci.* 22: 361–363.

Meyers, S., and Friedland, P. (1984). Knowledge-based simulation of genetic regulation in bacteriophage lambda. *Nucleic Acids Res.* 12: 1–9.

Ni, T. C., and Savageau, M. A. (1996a). Application of biochemical systems theory to metabolism in human red blood cells. Signal propagation and accuracy of representation. *J. Biol. Chem.* 271 (14): 7927–7941.

Ni, T. C., and Savageau, M. A. (1996b). Model assessment and refinement using strategies from biochemical systems theory: Application to metabolism in human red blood cells. *J. Theor. Biol.* 179 (4): 329–368.

Novak, B., and Tyson, J. J. (1995). Quantitative analysis of a molecular model of mitotic control in fission yeast. *J. Theor. Biol.* 173: 283–305.

Palsson, B. O., Narang, A., and Joshi, A. (1989). Computer model of human erythrocyte metabolism. *Prog. Clin. Biol. Res.* 319: 133–150; discussion, 151–154.

Sauro, H. M. (1993). SCAMP: A general-purpose simulator and metabolic control analysis program. *Comput. Appl. Biosci.* 9: 441–450.

Tanaka, K. R., and Paglia, D. E. (1995). Pyruvate kinase and other enzymopathies of the erythrocyte. In *The Metabolic and Molecular Basis of Inherited Disease*, 7th ed. New York: McGraw-Hill, pp. 3485–3511.

Tomita, M., Hashimoto, K., Takahashi, K., Shimizu, T. S., Matsuzaki, Y., Miyoshi, F., Saito, K., Tanida, S., Yugi, K., Venter, J. C., and Hutchison, C. A. (1999). E-CELL: Software environment for whole cell simulation. *Bioinformatics* 15 (1): 72–84.

Tyson, J. J. (1991). Modeling the cell division cycle: cdc2 and cyclin interactions. *Proc. Natl. Acad. Sci. USA* 88: 7328–7332.

SUGGESTED READING

Service, R. F. (1999). Complex systems. Exploring the systems of life. *Science* 284: 80–81, 83. General article on various aspects of systems biology, including computer simulation of the cell.

Tomita, M. (2001). Whole-cell simulation: A grand challenge of the 21st century. *Trends Biotechnol.* 19: 205–210. General article on cell simulation that focuses on whole-cell models and their potential practical applications.

URLs FOR RELEVANT SITES

Institute for Advanced Biosciences, Keio University.
http://www.bioinfo.sfc.keio.ac.jp/IAB

E-CELL Project. http://www.e-cell.org

Glossary

Annotation The process of associating knowledge to macromolecule sequences, usually genomes. This includes assigning all encoded protein activities, among them the predicted enzymes of a genome.

Apoptosis (also termed "programmed cell death," PCD) Regulated, active, morphologically and biochemically distinct process during development and tissue homeostasis in which a cell dies without causing inflammatory reaction in the surrounding tissue.

Autapomorphy A derived character state (apomorphy) that is unique to a particular species or lineage in the group under consideration. For example, within a set of taxa including fish, turtles, birds, and mammals, hair is a unique character, or autapomorphy, of mammals.

Boltzmann distribution The probability distribution at equilibrium for the occupancy of various states, each with a particular energy.

Cellular function The context of a protein; its interactions with other proteins in the cell.

Chemotaxis The orientation of an organism in relation to the presence of a particular chemical. The chemotactic response of *E. coli*, for example, depends on the ability to modulate their flagellar motor in response to external stimuli.

Cholesterol An amphipathic lipid that is an essential structural component of the cell membrane and outer layer of lipoproteins of blood plasma.

Clustering A process of grouping objects into classes of similar objects.

CluSTr A database that offers an automatic classification of SWISS-PROT and TrEMBL proteins into groups of related sequences.

CORBA Common Object Request Broker Architecture. An open vendor-independent architecture and infrastructure that computers use to work together over networks. It uses the Interchange Interorb protocol for communication in the operation between CORBA programs.

DBMS Database Management System. It is a computerized record-keeping system that stores, maintains, and provides access to information. A database system involves four major components: data, hardware, software, and users. The main advantage of using a

DBMS is that the formalism of the data model underlying the DBMS is imposed upon the data set to yield a logical and structured organization. Given a fuzzy, real-world data set, when a model's formalism is imposed on that data set, the result is easier to manage, define, and manipulate. Different data models lead to different organizations. In general the relational model is the most popular because it is the most abstract and the easiest to apply to data while still being powerful.

Differentiation Process or state in multicellular organisms (although one can also envision differentiation in bacteria) in which a cell acquires or maintains its functional specialization. Differentiated cells exhibit specific cell-type features and frequently lose their proliferative capacity.

Domains Protein domains are the separate and independent folding units of a protein. In sequence analysis these are in most cases consecutive stretches of sequence carrying a specific function, separated by small stretches of sequence with biased amino acid composition (low complexity regions). Enzymatic function can be better assigned if the different domains of the complete protein are considered.

Elementary mode Minimal set of enzymes that can operate at steady state with all irreversible reactions proceeding in the appropriate direction.

Entrez A retrieval system for searching several linked databases (`http://www.ncbi.nlm.nih.gov/Entrez/`).

Erythropoietin (EPO) A glycoprotein growth factor that regulates red cell production and adapts the red cell mass to the oxygen needs of the tissues.

Eukaryote An organism formed by cells containing a nucleus that has one or several chromosomes. Eukaryotic cells have a compartmentalized internal structure in which different cellular functions are carried out in membrane-bounded organelles, such as mitochondria, chloroplasts, endoplasmic reticulum, and Golgi apparatus. All multicellular, and many single-cell, organisms are eukaryotes; bacteria are prokaryotes (cells with no nucleus).

Expression array Method using DNA chips to simultaneously measure the mRNA expression levels of a large number of genes, usually all genes for a given organism.

External metabolites Metabolites that are buffered as they are connected to a reservoir. Nutrients would be sources; waste products would be sinks.

Feedback Feedback occurs when the output of a process influences its own subsquent states.

Formal language A formal language L over the alphabet A is a subset of A*, where * denotes the star operator. Production systems are used to describe formal languages.

Gene network The molecular genetic system that controls the processes occurring in the organism on the basis of hereditary information contained in its genome.

Genpept A database of translated protein-coding sequences derived from entries in the GenBank nucleotide sequence data bank.

Glycolysis Primary energy metabolism in cytosol in which sugars are degraded and ATP is produced.

Growth, growth factors   In microbiology the term "growth" is used to describe the process by which a single-cell organism increases its size (cell volume). Often, cell growth is coupled to cell division, in that cells double their size before undergoing mitosis. In contrast, in conjunction with mammalian biology, "cell growth" is used interchangeably with cell proliferation (i.e., the process of increasing cell numbers and tissue mass by cell division. "Growth factors" are natural, mostly proteinaceous molecules that bind with high specificity to cellular targets and stimulate cell division. "Growth factors" are often used interchangeably with "cytokines," which comprise a large family of proteins that have diverse effects on cells, including the stimulation of proliferation, migration, and differentiation.

Homology   From the Greek for "study of sameness." In biology, "homology" refers to similarities that are due to common ancestors having possessed the similar feature. Homologies are the basis of inferences about the nested relationships among taxa and are commonly used to motivate borrowing annotations in one biological context and applying them in another.

HSSP   A database of Homology-derived Secondary Structure of Proteins.

IDL   Interface Definition Language is the language for describing interfaces of CORBA programs. For every CORBA implementation (binding), there is a precompiler to produce client and server interfaces for a given document written in IDL.

Internal metabolites   Metabolites participate only in reactions that are explicitly taken into account in the pathway model. The formation of each internal metabolite has to be exactly balanced by its consumption (steady-state assumption).

InterPro   An integrated documentation resource for protein families, domains, and sites (`http://www.ebi.ac.uk/interpro/`).

Irreversible reaction   A reaction in which the rate of the forward reaction is so much higher than the rate of the reversible reaction that the latter is relatively negligible.

IS   An information system (IS) describes a coupling between a data storage system and further data-processing applications. Such an IS can be characterized by its main functions: data storage, information retrieval, data interconnection, and analysis of information.

JAVA   JAVA (TM) is a programming language designed for the development of computer platform–independent software. To execute programmed applications in JAVA, an interpreter (virtual machine) is required. JAVA programs are frequently used to implement Web pages.

JDBC   The JDBC (TM) technology is an application programming interface that permits access to any relational data source from the JAVA (TM) programming language. It provides cross-DBMS connectivity to a wide range of SQL databases.

Kinase   An enzyme that catalyzes the transfer of a phosphate group from ATP or other nucleoside triphosphate to a substrate.

Kullbach-Leibler distance   Also known as relative entropy, this is a measure of the dissimilarity between two probability distributions. The distance is 0 for identical distributions and increases as the distributions diverge.

Markov chain A series of states in which the probability of a state depends on previous states. In a zero-order Markov chain, the states are independent. In a first-order Markov chain, the probability of the current state depends on the previous state. In a second-order Markov chain the probability of the current state depends on the previous two states, and so on.

Metric A quantitative measure of similarity or dissimilarity.

Molecular function The specific biochemical function of a protein: its catalytic, structural, or binding activities.

Negative feedback This occurs when the output of a process weakens its intensiveness. Negative feedback can maintain automatic stability (homeostasis) of the regulated parameters at the level required.

Operon A transcribed polycistronic unit (group of genes) with its associated regulatory sites. An operon has to have at least two cotranscribed genes, as opposed to a transcription unit, which can be a monocystronic unit. The definition used here implies that a gene can belong only to a single operon but to several transcription units. Operons are often found in prokaryotic genomes, and the genes within an operon are often functionally related. See *Transcription unit*.

OQL Object Query Language is a query language comparable with SQL that supports the data model of the Object Database Management Group.

ORF The Open Reading Frame is a continuous region of DNA coding for a protein, or a fraction of it, found in the genome. ORF is usually distinguished from a gene in that the ORF is a predicted gene based on computer analyses.

Orthologous genes Two homologous genes in two different organisms that have evolved from a common ancestor.

Orthology Homology of genes between lineages (contrast paralogy).

Palindromicity A measure of the similarity of a sequence element to its reverse complement.

Paralogous genes Two homologous genes that diverged from a duplication event in the same organism.

Paralogy Homology in which a gene duplication allows related proteins to evolve independently within the same lineage (contrast orthology). Paralogues can be found in the same individual. The orthologue/paralogue distinction is relevant when trying to compare proteins across lineages.

PDB The Protein Data Bank is a repository for the processing and distribution of 3-D macromolecular structure data primarily determined experimentally by X-ray crystallography and NMR.

Pfam A collection of protein family sequence alignments that are constructed semi-automatically, using hidden Markov models.

PHP PHP, which stands for "PHP: Hypertext Preprocessor," is an HTML-embedded scripting language. Much of its syntax is borrowed from C, JAVA, and Perl with some unique PHP-specific features included. The goal of the language is to allow Web developers to write dynamically generated pages quickly.

Phylogenetic profile A vector constructed for a given gene in a particular organism, representing whether an orthologue to a particular protein is present or absent in each additional genome sequence. It forms the basis for a computational method of predicting protein function.

Positive feedback This occurs when the output of a process enhances its own activities. Positive feedback leads to the system's deviation from its stationary state.

PRINTS A compendium of protein fingerprints. A fingerprint is a group of conserved motifs used to characterize a protein family.

ProDom A protein domain database that consists of an automatic compilation of homologous domains.

Production system A production system is defined by a finite set of production rules and a finite alphabet A. Production rules are pairs of words over the alphabet A* that define the derivation of the production system. A word w over A* will be derived by a production rule (u, v) if the first component of the production rule (u), which will be a subword of w, will be replaced by v. The rule is to be read as "rewrite u by v."

Prokaryote An organism, usually single-celled, formed by cells with no nucleus; it has a simple internal organization lacking organelles.

PROSITE A database of biologically significant sites and patterns formulated in such a way that, with appropriate computational tools, it can rapidly and reliably identify to which known protein family, if any, the new sequence belongs.

Protease An enzyme that degrades proteins by hydrolyzing peptide bonds between amino acid residues.

Protein fusion A protein containing two or more domains that appear separated in other instances.

Protein kinase Enzyme that catalyzes phosphorylation of (transfers of a phosphate group from ATP to) the hydroxyl groups of tyrosine, serine, and threonine residues of proteins. Phosphorylation of proteins often regulates their enzymatic activity or their affinity to other proteins. Signal transduction events typically consist of protein phosphorylation reactions in which protein kinases act both as substrates and as enzymes whose kinase activity itself is induced by phosphorylation, thereby forming a cascade.

Proteome The complete set of proteins encoded by a genome.

QBE This nonprocedural language developed by IBM(TM) was designed for textual querying and manipulating of relational DBS. QBE means "Query by Example" and is implemented as an intuitive query construction using tables.

Regulatory motif (in DNA) A short conserved sequence element (usually 10–20 basepairs long) that is specifically recognized by a regulatory transcription factor.

Regulon A group of operons or genes controlled by one regulator. The initial definition was the result of studies of the ArgR regulon in *E. coli*. It is common to find a relaxed notion of a regulon as a group of operons or transcription units that are controlled by a common regulator, independently of additional regulators affecting each operon or transcription unit of the regulon differently.

Relative entropy Measures the difference between two probability distributions. Also known as Kullbach-Leibler distance.

Rosetta Stone protein A protein composed of two or more domains, each of which can be found as an independent protein; used to find functional links between the independent proteins.

RuleBase A database that stores and manages annotation rules derived from SWISS-PROT and InterPro for the automatic annotation of TrEMBL.

Semi-Thue System A Semi-Thue system is a specific production system that is equivalent to the Chomsky type-0 grammar.

Sequence search Comparison of a reading frame for significant (nonrandom) similarity against any or some of the reading frames stored in a large database, using a sequence similarity search algorithm.

Serial homology Repetitive structures of the same organism. The arms and legs of tetrapods, mammalian cervical vertebrae, and the leaves and the branches of a tree are all serial homologues. Crayfish have 19 pairs of appendages. These jointed appendages all reflect the same basic morphological pattern but serve different functions, from chewing to food handling to walking to mating. Serial homologies are not used to reconstruct phylogenies, but they are relevant to understanding development.

Sigma factor Bacterial proteins that interact with RNA polymerase and dictate promoter specificity.

SignalP A signal peptide prediction program.

Signal transduction pathway The process by which the information contained in an extracellular physical or chemical signal (e.g., a hormone or growth factor) is received at the cell by the activation of specific receptors, then conveyed across the plasma membrane and along an intracellular chain of signaling molecules, to stimulate the appropriate cellular response.

SPTR A comprehensive, nonredundant database composed of SWISS-PROT, TrEMBL, and TrEMBLnew sequences.

SQL Structured Query Language is a nonprocedural language for interacting with a relational DBMS. It allows the user to query, manipulate, and define data.

SRS Sequence Retrieval System is a specific system that incorporates tables from many different biological databases.

Stiffness problem The problem of stiffness occurs in a simulation system where there are very different scales of the time variable on which the dependent variables are changing.

Structured substance-reactor model Framework for object—oriented cell simulation used in the E-CELL system. It consists of three classes of objects, Substance, Reactor, and System, representing molecular species, reactions, and functional/physical compartments, respectively.

SWISS-PROT An annotated protein sequence database. It contains high-quality annotation, is nonredundant, and is cross-referenced to many other databases.

Symplesiomorphy Ancestral (or primitive) character state that is shared by two or more taxa. Shared ancestral character states are not helpful in grouping taxa when producing the nested clades in a phylogenetic tree.

Synapomorphy Derived character states (apomorphies) shared by two or more taxa (i.e., the features have a common derivation). If the two groups share a character state that is not the primitive one, it is plausible that they share a more recent common ancestor than the one shared by the out-group (which defines the ancestral state). Synapomorphic character states are used as evidence to group taxa into nested groups. Phylogenetic trees are built by grouping taxa united by synapomorphies.

TMHMM A transmembrane prediction program.

Topology of a metabolic network This is a set of enzymes and metabolites, their connections, and the stochiometry and directionality of the reactions.

Transcription factor Regulatory protein that binds specifically to short DNA sequences (5–20 nucleotides) and interacts with RNA polymerase enzyme in order to modulate transcription.

Transcription unit One or more genes that are cotranscribed. More precisely, we have limited its use to one or more cotranscribed genes from a single promoter. The main differences between a transcription unit and an operon as here defined are (1) operons must have at least two genes, and (2) a gene can belong to one or more transcription units but to only one operon. See *Operon*.

TrEMBL A computer-annotated supplement to SWISS-PROT. TrEMBL contains the translations of all coding sequences (CDS) present in the EMBL Nucleotide Sequence Database that are not yet integrated into SWISS-PROT.

XML Extensible Markup Language is a format for structured documents and data. It is well defined by the World Wide Web Consortium (W3C) and has become a standard for text-based data exchange.

Corresponding Authors

Jeremy C. Ahouse
Millennium Pharmaceuticals, Inc.
Cambridge, Massachusetts

Rolf Apweiler
EMBL Outstation—European
Bioinformatics Institute Wellcome
Trust
Hinxton, Cambridge, UK

Julio Collado-Vides
Centro de Investigación sobre
Fijación de Nitrógeno, UNAM
Cuernavaca, Morelos, México

Thomas Dandekar
Institute for Molecular Medicina
AG Biocomputing and Biological
Structures
Heidelberg, Germany

Ralf Hofestädt
Bioinformatics/Medical
Informatics
University of Bielefeld
Bielefeld, Germany

Sui Huang
Harvard Medical School
Surgery Research Laboratory
Children's Hospital Medical
Center
Boston, Massachusetts

Nikolay A. Kolchanov
Institute of Cytology and
Genetics SB RAS
Novosibirsk, Russia

Abigail Manson McGuire
Department of Genetics
Harvard Medical School
Boston, Massachusetts

Edward M. Marcotte
Department of Chemistry and
Biochemistry
Institute of Cell and Molecular
Biology
University of Texas at Austin
Austin, Texas

Gary D. Stormo
Department of Genetics
Washington University School of
Medicine
St. Louis, Missouri

Masaru Tomita
Institute for Advanced
Biosciences
Keio University
Fujisawa, Japan

Index

domain, 40–41
role, 160, 167–168
Lysine, 263
Lytic-lysogenic switch, 282–283

MACAW. *See* Multiple Alignment
  Construction and Analysis Workbench
Major histocompatibility domain, 40
Mammals, 172–173
MAP (mitogen-activated protein) kinase,
  185, 187, 193
cascades, 208
MAP score, 138
MARGBench. *See* Modeling and
  Animation of Regulatory Gene
  Networks
Markov models, hidden (HMMs), 23
Mating type, 142
Matrices. *See* Elementary modes; Weight
  matrices
MEME. *See* Multiple EM for Motif
  Elicitation
Metabolic flux space, 259
Metabolic networks
diagrams, 59
glucose-6-phosphate, 60–61
glycolysis pathway, 67–69
and homology, 223
MARGBench example, 56, 69–76
mathematical models, 58–59
Metab-Sim simulation, 58, 66–69, 71
in multicellular organisms, 182
rule-based modeling, 60–66, 76–79
simulators, 163–167
Metabolic pathways
alignment, 259–263, 265
boundaries, uncertain, 267
branched and interconnected, 252–253
catabolic and anabolic, 252
defined, 252–253
genome analysis, 263–267
identification
  elementary modes, 255–259, 263
  by synthesis, 255
and organism complexity, 268
Metabolism
comparative analysis, 41–42

proteome proportion, 41–42
Metabolites
identification, 257–258
in models, 60, 62–63, 65
MetabSim, 58, 66–69, 71
and MARGBench, 75–76
Methanococcus jannaschii, 263
Metrics, of protein function, 238
Mevalonate pathway, 159–160
MGD. *See* Mouse Genome Database
Michaelis-Menten kinetics, 66–67, 165, 167,
  193
Microarrays, 235
Millipedes, 5
Minimal upstream regions, 116
MIPS. *See* Munich Information Centre for
  Protein Sequences
Mitochondria, 283–285
Modeling and Animation of Regulatory
  Gene Networks, 56, 69–75
application, 75–76
Models. *See also* Boolean networks;
  Simulation
chemical-kinetic, 164–167
of cholesterol biosynthesis, 167–169
gene regulation, 65
mathematical, 150, 191–192
of metabolic pathways, 267–268 (*see also*
  Metabolic pathways)
rule-based, 60–66
Modularity, 190–191
Molecular databases, 53–58
Morphogenesis, 160–163
Motifs
common, 112
discovery process, 129, 130
DNA-binding, 114
DNA regulatory, 135–140
overlapping, 112–113
palindromic, 140
RNA-binding, 38
Mouse Genome Database (MGD), 33
mRNA expression, 130–131, 142–143, 235–
  236
Multiagents, 57
Multicellular organisms, 181–182, 209–
  212

homology, 223
modularity, 190–191
perturbations, 189–190
simulation, 280–282
subcellular, 212–213
Similarities
explanations, 6–8
nonhomologous, 5
in phylogenetic profiles, 229
statistical significance, 28–29, 265
Simulation. *See also* Models
of cell (*see* E-CELL)
chemical kinetic, 164–167
E. coli chemotaxis, 281
essential data, 285–286
metabolic pathways, 268
of networks, 163–167
signal transduction, 191–192, 214
software for, 275–277
stiffness problem, 276
Simulators, 49–50, 58
Size, of proteins, 36–37
Sodium transport query, 30–32
Software, simulation, 275–277
Specificity
definition, 109
of DNA-binding protein
quantitative aspect, 87–89
weight matrix models, 89–97
RNAP transcription, 106–107, 110–117
to species, 263
Specificity score, 138–139
SPTR database, 26, 28
SQL. *See* Structured Query Language
SREBP proteins, 155, 159–160, 168
Stanford Microarray Database, 235
State space, 197, 198–200, 203, 214
Statistical analysis, 27, 31
Statistical significance
of similarities, 28–29
Sterol-regulated protease, 155–156, 160
Stiffness, 276
Structure-function relationships, 266
Structured Query Language (SQL), 73, 75
SWISS-PROT
annotation, 21–23, 24

and clustering, 28
and GO, 34
and protein functions, 238, 266
and proteome analysis, 26
controlled vocabulary, 33
database, 54
Synchronous congruence, 119
SYSTERS (SYSTEmatic Re-Searching), 28

T. pallidum, 266
Thermotoga maritima, 266
TIGR. (The Institute for Genomic
Research), 261, 264
TIM barrels, 266–267
Tissue dynamics, 210
Tissues, morphogenesis, 160–163
TMHMM, 27, 37
Transcriptional units (TUs)
in bacteria, 116–117
boundary genes, 116, 121, 122, 124
Transcription factors
binding sites, 94, 98, 129
and Boolean networks, 208, 213
and cholesterol, 155, 159
and cis-regulators, 213
of erythrocyte differentiation, 162
Gene-Net depiction, 154, 155
for purine biosynthesis, 132
site recognition, 234
Transcription regulation
activators *vs.* suppressors, 114–115
in bacteria, 114
evolutionary origin, 105, 121
false positives, 106, 109, 111–112
and gene function, 114
initiation, 104, 106
Transcription Regulatory Regions
Database (TRRD), 54
Transcriptomes, 2
TRANSFAC database, 54
Transition complexes, 193
Transmembrane proteins, 27
TRANSPATH, 49
TrEMBL
annotation, 21–23, 24
and clustering, 28

controlled vocabulary, 33
  and GO, 34
  and proteome analysis, 26
TRRD. *See* Transcription Regulatory
    Regions Database
Tus. *See* Transcriptional Units

Von Willebrand factor, 36

WD repeat, 38
Weight matrices
  and DNA regulatory motifs, 139
  of *E. coli* promoter, 107, 108–110, 113
  expectation-maximization (EM), 96
  Gibbs' sampling, 96–97
  greedy algorithms, 95
  limitations, 110
  network method, 97
  in RegulonDB, 111
  as specificity model, 90–93
Wings, 4, 5
WIT database, 54, 140–141, 231, 261

Yeast. *See also Saccharomyces cerevisiae*
  and domain fusion, 227
  in network, 241–242, 243
  phylogenetic profile, 230
Yeast Protein Database (YPD), 135, 142